melted away

melted away

A MEMOIR OF CLIMATE CHANGE & CAREGIVING IN PERU

Barbara Drake-Vera

LOUISIANA STATE UNIVERSITY PRESS

BATON ROUGE

Published by Louisiana State University Press
lsupress.org

This work depicts actual events in the life of the author as truthfully as her recollection permits and/
or can be verified by research. Occasionally, dialogue consistent with the character speaking
has been supplemented. All persons within are actual individuals; there are no composite characters.
The names and identifying characteristics of some individuals have been changed to respect
their privacy, as have the names of certain health care facilities.

LSU Press Paperback Original

DESIGNER: Michelle A. Neustrom
TYPEFACES: Whitman, text; Bicyclette, display

Maps created by Mary Lee Eggart.

COVER IMAGES: Photograph of Mount Ausangate by Jorge Vera, 2008.
Traditional weaving from Cusco region of Peru.

LIBRARY OF CONGRESS CATALOGING-IN-PUBLICATION DATA

Names: Drake-Vera, Barbara, author.
Title: Melted away : a memoir of climate change and caregiving in Peru / Barbara Drake-Vera.
Other titles: Memoir of climate change and caregiving in Peru
Description: Baton Rouge : Louisiana State University Press, [2024] | Includes bibliographical
    references.
Identifiers: LCCN 2023035241 (print) | LCCN 2023035242 (ebook) | ISBN 978-0-8071-8152-2
    (paperback) | ISBN 978-0-8071-8192-8 (pdf) | ISBN 978-0-8071-8191-1 (epub)
Subjects: LCSH: Drake-Vera, Barbara. | Drake-Vera, Barbara—Family | Americans—Peru—
    Biography. | Women authors, American—Biography. | Women caregivers—Biography. |
    Climatic changes—Peru. | Alzheimer's disease—Patients—Care—Peru. | Peru—Environmental
    conditions—21st century. | Peru—Social conditions—1968–
Classification: LCC F3448.7.D73 A3 2024 (print) | LCC F3448.7.D73 (ebook) | DDC 985/.004130092
    [B]—dc23/eng/20240102
LC record available at https://lccn.loc.gov/2023035241
LC ebook record available at https://lccn.loc.gov/2023035242

for Jorge, with all my love

in memory of John Drake and Isabel "Chata" Du Bois Hale

Jealousy can open the blood, it can make black roses.

—SYLVIA PLATH, "THE SWARM"

solastalgia: the psychic pain experienced when a familiar place
is irrevocably altered by the effects of climate change; "the homesickness
you have when you are still at home."

—TERM COINED BY AUSTRALIAN PHILOSOPHER GLENN ALBRECHT
IN 2003; DERIVED FROM THE LATIN *SOLARI* (COMFORT IN THE FACE OF
DISTRESSING FORCES) AND GREEK *ALGIA* (PAIN, SUFFERING)

CONTENTS

ix

ILLUSTRATIONS

melted
away

Embodying a figure from Andean mythology (a powerful being who is half-man, half-bear), a young *ukuku* carries a large candle to the top of Qolqepunku Glacier at the Qoyllur Rit'i pilgrimage, June 2006. Photo by Jorge Vera.

PROLOGUE

June 14, 2006, Qoyllur Rit'i (Snow Star) pilgrimage site, Sinakara Valley,
Cordillera Vilcanota, fifty miles east of Cusco

I gasp for air, my head throbbing in pain, as I trudge behind Jorge—my photographer, my husband, my partner in wild, improbable leaps of faith—up the
muddy, stone-strewn path to Qolqepunku, a tropical glacier high in the southern Peruvian Andes. Above us, the glacier's white mass towers between two
sharp black peaks; ahead on the trail march six young *ukukus*—guardians of the
glacier—men in long, shaggy black robes adorned with red crosses and tinkling
brass bells. They represent mythical bear-men from Andean cosmology and are
said to battle the condemned souls who haunt the glacier at night. Beside me is
our local guide, Paco, bright-red pom-poms dangling from his brown felt hat.
He speaks mainly Quechua, an indigenous language of the highlands, and a
hit-or-miss Spanish that is still better than mine.

A thousand feet below us lies the Catholic sanctuary and a wide, treeless
basin dotted with camping tents and around one hundred thousand pilgrims
from throughout the Andes. They have climbed nearly to the top of this eighteen thousand–foot–high mountain, as people have since pre-Inca times, to
sing and dance and pray for a good harvest and to pay homage to the Great
One: El Señor de Qoyllur Rit'i, Lord of the Shining Snow Star, a divine entity
who is either Christ or the living spirit of the mountain itself, depending on
which priest, shaman, or anthropologist you talk to.

I have never been this high in the Andes before, not even when Jorge and
I married ourselves at Machu Picchu a decade ago. I have never seen a glacier
up close. I am here because Jorge decided he had to photograph a precarious

1

Modern Peru, with locations visited by the author between 2006 and 2012 and featured in *Melted Away*.

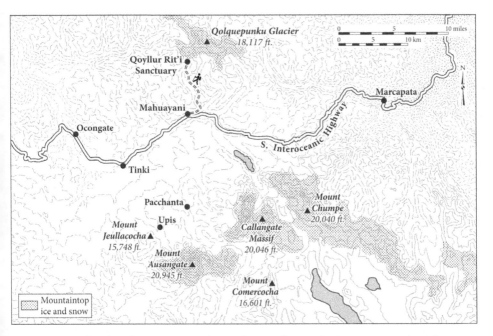

The Qoyllur Rit'i pilgrimage site in the Sinakara Valley, Mount Ausangate, and nearby sites in Peru's Cordillera Vilcanota. The dotted vertical line indicates the route that pilgrims climb on foot from the town of Mahuayani to the Qoyllur Rit'i sanctuary, which is maintained by a lay Catholic brotherhood. Pilgrims pay annual homage to a Christian lord (El Señor de Qoyllur Rit'i) as well as to the mountain deity Apu Ausangate, whose seat lies due south of the pilgrimage sanctuary. These two sites coexist within an ancient system of sacred geometry (ceques) relative to Cusco, the former seat of the Inca Empire, notes Michael J. Sallnow.

ice ritual performed by a bunch of daredevil *ukukus* half his age, and I want to make sure he does not fall into a crevasse and die.

Thom-thom-THOM-da-dom, beat the drums. *Ching, ching, ching,* ring the *ukukus'* bells.

I draw short, jagged breaths, head pounding, and force myself to lift one foot at a time. Every step is a torture. I want to go back to the campsite. I didn't realize forty-five is old.

"Every year the *ukukus* run down the mountain with holy ice for the cathedral," Paco is saying to Jorge, who translates for me. "To bless the crops and the animals and the people for the year."

Jorge nods in profile; he has a beautiful strong nose whose high bridge is always cool to the touch. Right now, that nose is practically all you can see of him, we're both so bundled in scarves and knitted llama wool caps and poufy down jackets.

"Don't the *ukukus* sometimes slip on the ice and die?" Jorge asks.

"Yes, some die every year," Paco says in his raspy voice. "But the death of an *ukuku* blesses everyone."

At the blast of a whistle, the young *ukukus* sprint to the base of the glacier, about two hundred feet away. *Ukukus* from other villages, some in green-and-black robes, are gathering there too, carrying a large wooden cross and three-foot-long wax candles. According to ancient tradition, the *ukukus* are supposed to camp out here tonight, carve out chunks of ice, strap them to their backs, and race down the mountain early tomorrow—however, today something is wrong, apparently; they're arguing.

"What are they saying?" Jorge asks Paco. "Can I photograph them tonight?"

Paco races up the path to check. He has lived at this altitude all his life, so breathing isn't a problem for him.

"No photographs, no ice camping," Paco explains when he returns, his thick eyebrows pinched in worry. The ice is no longer stable, he says. It's too risky. Only a small group of *ukukus* will do the vigil on the glacier tonight. And no race at dawn down the mountain with the holy ice. Too many *ukukus* have died in the last few years. Big holes have opened up in the ice.

Jorge grows very still. I can taste his disappointment. We flew from Florida for this.

"Why are there big holes?" I ask Paco.

"The glacier is dying. Climate change, of course." *El cambio climático, por supuesto.*

I am stunned. I had heard the high-altitude tropical glaciers in Peru were melting, but I didn't realize this one was disappearing too. And so quickly.

Por supuesto. That was a dig. He's a llama herder from the Andes, and he knows the science behind the melting glaciers. Why don't I, an American tourist in expensive hiking boots, know these basic facts, he insinuates? Meanwhile, in the United States, news outlets are debating whether climate change is "real," and right-wing politicians are decrying global warming as a "hoax."

If only I could bring those people here, show them climate change is real, I think.

Squinting up into the sun, I see the face of the glacier is crawling with people—gathering ice, laughing, throwing snowballs at each other. A young boy lifts an impressive chunk of ice over his head like a trophy. "¡El Señor de Qoyllur Rit'i!" he shouts. Meanwhile, Jorge is trying to convince the *ukukus* to let him photograph their rituals. It's three o'clock already. By four, the sun will have disappeared behind the mountains, and temperatures will quickly plummet below zero. The dangerous time. The time when cocky gringos die. Like the experienced French climber who insisted on hiking Qolqepunku yesterday at nightfall wearing only shorts and a T-shirt and got hypothermia. They brought his lifeless body down on a cloth stretcher.

Fuck it, I think, ignoring my headache. *I'm going up while there's still sunlight.*

I grab my hiking poles and haul myself up the moraine path, shards of dark-gray slate crunching underfoot. Every few steps, I have to stop to catch my breath. My head throbs.

Up close, the glacier is nothing like I imagined. Its walls are like the flanks of a whale, pale-gray and wet and glistening, its surface studded with grit. The glacier feels ancient, alive—and sentient. I want to close my eyes and lean against it, let my humanity dissolve in the trickle of icy water that seeps from underneath, joining the other rivulets winding to the valley below.

A deep vertical crack has split the ice face to my left. Very carefully, I creep along the edge of the glacier. The color inside takes my breath away. It is a translucent blue-green, lit from within, a beauty that pierces my heart. I stand there, forehead resting on the gritty ice, filled with something improbable. Affection for a glacier. It reminds me of something a shy cabdriver told me a few days ago in Cusco: "Yes, we love our mountains and their white ponchos." White ponchos. He was referring to the ice caps.

Apu Ausangate, the name the local people call the deity who protects this particular mountain range. Lord Ausangate. The living spirit of the mountains. This *apu* is male, like most Peruvian mountain spirits. He is a good *apu*, a benevolent father, his glacier meltwater giving life to the people of this high, forgotten region. "Children of the glacier," they call themselves, a Peruvian scholar recently wrote.

"The pilgrimage of Qoyllur Rit'i is when they climb the glacier to play in the lap of their father," she wrote. "He will always be there for them, until the snowcaps disappear."

What must that feel like, to be nurtured and sustained by one's father? I wonder, gazing into the blue ice. *To be loved completely, no conditions demanded in return?*

I try to imagine it, but I can't. *Father* and *love* and *nurture* don't go together in my brain.

The ice face is turning purple; it must be time to go. I grab my hiking poles and retreat backward, inch by painful inch. At the trailhead, I double over, my head pounding so hard I want to puke. Altitude sickness.

Pebbles rain down to my right. The pilgrims are climbing off the glacier and returning to their campsites. Whole families stream by, bumping into me, laden with buckets of meltwater and blocks of ice. A man in a red poncho steps on my toe, and I don't even feel it.

Then something is tugging on my sleeve—rough fingers, the nails caked with black dirt. It's Paco. He holds out a dripping chip of glacier ice, two inches long.

"Señora, for you." His raspy voice.

I thank him and let the fifty thousand–year–old ice melt on my tongue, down my throat, the iciness radiating throughout my chest, strangely cold and warm at the same time. The feeling lodges in my heart. For some reason, I don't mind, just like I didn't care about washing the gravel from the ice when he gave it to me. I stand there, quietly, just looking at Paco and his thick, slanty eyebrows and his chapped red cheeks. His chest rises and falls under his worn fleece jacket, puffs of frosty air blooming between us.

Then I notice: my headache is gone.

"Gracias," I say. He dashes off to gather more ice.

"Barbara, we have to get going," Jorge is calling down below.

The glacier is bathed in purple shadows. Over my right shoulder, the sun is sinking fiery orange between the mountain peaks.

I glance up to the top of the glacier. *I will write an article about you,* I promise silently. *About you and Qoyllur Rit'i and climate change, and get it published, if it's the last thing I do.*

But who is going to care what you say? an old, mocking voice whispers inside. *You're no science writer. You know jack shit about climate change.*

My head begins to throb.

Who in the United States is going to listen to you?

I hurry down the path, rock crunching underfoot. The cold air whips my eyes and face. I'm walking faster, faster, almost running, then I stumble; my left pole catches me before I pitch forward. I keep going. I'm sweating into my chilled jacket, my heart pounding, pounding.

The voice chases me as I rush into darkness: *Face it, Barbie. You haven't got it in you.*

That was 2006, a year before Jorge and I fled to Peru permanently, two years before the Great Recession brought the U.S. economy to its knees, a decade before many Americans began acknowledging that the extreme weather events—monster storms, heat waves, severe floods, drought—that have been battering our country for decades are due to the effects of human-induced climate change.

Four years and seven months after I fell in love with a glacier, my eighty-six-year-old father, John Drake, a veteran of Pearl Harbor, was diagnosed with Alzheimer's. Two months later, with Jorge's help, I became his reluctant care-giver in the desert capital of Lima, Pizarro's "City of Kings."

A year and a half later, my father drew his last breath—in bed with four young women.

The following month, in a small storage unit in Florida, I found a crumbling leather-bound journal that he had kept hidden for sixty years—and everything I thought I knew about him, and our long, painful history together, was blown wide open.

The author, age four, and her father, John Drake, forty-one, on the beach at Asbury Park, NJ, 1965. Photo by Ann Drake.

1

_____

FIRST THINGS

The Big Blue Table

My father was always jealous of me. Well, nearly always.

I first became aware something wasn't quite right when I was around nine years old. One night I was in our dining room in Bloomfield, New Jersey, writing at the long blue-gray table in the center of the room. I had sat here almost every day after school for the last two years, writing poems, calling to my mother for help with spellings. Around six o'clock, after returning from his job at a nearby post office, my father would drop by for a quick kiss on my head before retreating into the pantry for a "little pick-me-up," as he called it. On the wall to my left—above the bookshelves crammed with Modern Library editions and my father's collection of classical records—was an enormous, yellowing map of Paris, where he and my mother had lived in the early 1950s. To my right, tall windows mirrored my reflection against the deepening evening sky—a sturdy girl in tortoiseshell glasses and green pajamas, my long red hair bound in two ponytails, the elastics capped by hard, translucent gumballs.

An only child, I loved doing this, losing myself in word-sounds, writing short rhyming verses about birds and nature and Martin Luther King Jr. and freedom with a capital *F*. It was as everyday and wondrous as flying down the nearby hill on my roller skates. Only lately, thanks to the efforts of my third-grade teacher, my poems were being published in national children's magazines. And I was starting to receive fan mail, along with requests for more poems.

That evening—my mother now allowed me to stay up until ten o'clock to answer my mail—I was writing a poem for Miss Angle's fourth-grade class in

Rosedale, Illinois. My lucky Snoopy pen flew over the blue-lined paper as I tried to conjure up a short, stubby tree (*stubby, chubby*) with red mushrooms (no, *orange mushrooms*) at its base.

"How do you spell *squished?*" I yelled over my shoulder.

"S-Q-U-I-S-H-E-D," replied my mother from the living room, stifling a yawn. The springs of her armchair creaked as she stood to go upstairs to bed.

A while later, my father entered the dining room and picked up Miss Angle's letter. I ignored him and kept writing. He began circling the table, around and around, counterclockwise. No kiss on the head. Whatever. The letter fluttered to the table, by my elbow.

> But in the fall, his leaves fell down.
> His toadstools squished into the ground.

I smelled his boozy, tobacco-y scent. He was leaning over my shoulder. From the corner of my eye, I saw his face. Not a smile. An angry line. Like the blue Seine frowning over Paris.

He poked the page with his long, square-tipped finger. "*Squished?* Is that a word?"

"Yeah, like *squashed* but with a *squish* sound," I said. "Like when you step on a wet, slippery mushroom."

"Hmmmph," he said and muttered something that began with *anna*.

"What?"

"Oh, just forget it." His nasal voice went down in an ugly way on *-get it.*

"No, Dad. I want to know, what's an *anna-motta?*"

I looked up at him. His eyes were bloodshot, and his auburn hair, usually raked over his forehead in an elaborate comb-over, was dangling off one ear.

"Squish squash squish," he said. He laughed and disappeared into the pantry. I heard ice clinking in a glass.

I looked down at the page, a lump in my throat. *Squished.* What was wrong with that?

— ✦ —

Another indelible memory of my father—so painful it plagued me for years, setting off a wave of peculiar, shameful neuroses—dates to the spring of 1983. I was twenty-two years old and a student at Purchase College, where I was majoring in music. Since the age of twelve—after conflicts with my father over my getting published came to a head and I quit writing altogether—I had dedicated myself to becoming a professional oboist, practicing four or five hours a day, performing with orchestras and chamber groups, appearing on stages in New England and New York and France. All that time, my father had existed in a perpetual state of engorged pride, sitting front row at my concerts, snapping rolls of pictures, boasting of me to relatives and strangers, buying me anything I wanted—a Lorée oboe, concert dresses from Paris, a cream-colored Camaro with a gold stripe down the side. Nice things, beautiful things, my secret weakness. Only lately, at college, my need to write had resurfaced—so forcefully I could barely control it—and I was fed up pretending I wanted to spend the rest of my life bleating away in a symphony orchestra. I had to tell my father the news face to face, and here he was in the Purchase parking lot, dressed up in a three-piece suit and burgundy silk tie, eager to hear me play Respighi's *Pines of Rome* that night—and anxious to show off his brand-new Honda Accord.

I sat next to him on the front seat, the intoxicating new car smell shimmering all around us, as he rattled on about gas mileage and Dolby surround sound.

"I'm switching my major to English," I finally blurted out. "I want to become a writer."

His face turned beet red above the white shirt collar, lips clamped in a downward line. The angry Seine.

"What have you written so far?" he asked in a pinched voice.

"A couple of short stories. I'm just getting started, but—"

"A couple of stories?" he exploded. "A couple of stories? Face it, Barbie. You haven't got it in you. The stuff isn't pouring out of you!"

"Yeah, but my writing teacher says—"

"You're deluded!"

He kept ranting, his nasal voice getting higher and louder. It would take me ten years to develop a good prose style and then what? I would be competing with thousands of other writers who wanted to do the same thing: write the

Great American Novel. I'd be poor and miserable, a failure, stuck in a garret with a pile of rejection slips.

I stopped listening. All I could see was his blood-red face and the angry word-sounds twisting his mouth. The man who had written a single novel in his twenties, in Paris, and spent the next forty years as a worker bee, first as a market researcher in New York and then as logistics specialist for the U.S. Postal System. Who was he to give me advice?

I gulped air through my mouth. It was hard to breathe.

"Stop kidding yourself, Barbie," he was saying. "You'll never make it."

I wrenched the car door open and ran to my practice room to write it all down.

Some parts quickly faded from memory. But a few phrases struck with perfect accuracy. They lodged in some soft, crenulated place open only to him, Daddy Dearest. Each time I thought back to the scene in the car, the words sank deeper, filling me with dread.

Face it, Barbie, you haven't got it in you. The stuff isn't pouring out of you.

Soon I didn't have to try to remember. The words spoke themselves. Like a curse.

I would be in the shower, water dripping on my head, and I would hear his pinched, nasal voice, the words echoing inside me, pricking a sense of helplessness and shame that flushed through me, down my neck, chest, arms, down, down, down. Shame that I cared. Shame that I was weak. Shame that he could get to me like this and I was powerless to stop it.

Stop caring, I told my crying self. *You're a writer, right? Stop fucking caring. Stop fucking caring.*

A couple of weeks later, I changed my major. A kind-faced woman at the registrar's office mistakenly credited my music composition classes as courses in English composition, so I only had one more year of coursework to earn my degree in English. At least the universe was cooperating, I told myself.

The following spring, when I moved to Manhattan to find a job in publishing, my father sold my Camaro and cut me off with eighty dollars and twenty-three cents in the bank. It was all the money I had.

"Good luck," he said over the phone. "It's a cruel world out there."

"Your father's always been a bit out of it," my mother said, lighting a cigarette on the deck outside their home in Toms River. This was March 1991, and by that time, I had been editing a music magazine for five years—grossly underpaid but buoyed by secret infusions of cash from my mother every few months, mailed to me by check or shoved at me, in crisp twenties, from her twisted, arthritic hand on the rare occasions we saw each other. Two months after my father disowned me, she sent two thousand dollars so I could rent my first apartment, a fourth-floor walkup in Washington Heights, saving me from ruin. Just as important were her weekly letters, urging me not to stop writing and to keep on my path. "No bird soars too high if she soars with her own wings," she wrote in emphatic, if wobbly, downward strokes, tweaking Byron's words.

I gazed at her now as she puffed on the cigarette, her once-brown hair suffused with gray, her cheeks gone droopy, her keen brown eyes magnified by thick glasses. Since contracting rheumatoid arthritis in her forties, life hadn't been easy for her, but she was stoic and didn't complain or let it stop her from doing things. She just kept hobbling around on her knobby feet and writing her letters by hand and cleaning their two-story house and making meals for her and my father, who really should have noticed she could have used some help. Right at this moment, I could see him on the other side of the sliding glass doors, inside the kitchen, meticulously buttering a slice of toast. He didn't bother looking up: After eight years, I was still to be held at arm's length, an investment gone bad, a not-unspoken disappointment. When I came to their home, I did so for her. He refused to admit our conversation at Purchase had happened or that he'd left me penniless in New York. He considered these matters "solved" a few years ago when he dumped off a new dhurrie rug, unasked for, at my apartment. I could still recall his tense face sagging with relief when he got back in the car and drove away over the potholes.

"You keep saying he's 'out of it,'" I said to my mother. "What has he done this time?"

"Oh, forgetting things. And he won't stop ordering vitamins."

"That doesn't sound too bad."

"Just look in the guest room closet."

That afternoon, I tugged open the folding doors. The small closet had been fitted floor to ceiling with wire shelves. Every inch of shelf space was taken up with family-size bottles ordered from Pilgrim's Pride—vitamin E, vitamin C, glucosamine, ginkgo biloba. I counted at least 150 unopened bottles. On the bottom shelves were boxes of Nice 'N Easy auburn hair dye and cans of Final Net. One whole shelf was filled with hard-cover biographies of American presidents and the Masons and Sir Francis Drake, the English pirate whose brother Robert, my father insisted (without proof), was our long-ago ancestor.

"So, he's watching his health and maybe yours," I told my mother later outside. She was leaning on the wooden railing, looking at the scrubby lawn.

"It's not just that," she said, face flinching. "It's other things."

"Like . . . ?"

"It doesn't matter."

"Yes, it does, Mom. I worry about you."

She took my hand and stroked it. "I'll be okay. He's always been a bit out of it. You know that."

I wrapped my arms around her body and hugged her. She was so shrunken now, my head rested on top of hers. It smelled reassuringly of her.

"Whatever you do," she said fiercely, "don't stay here because of me."

She glared up at me. The rims of her eyes were red, but there were no tears. The rheumatoid arthritis had taken all the crying out of her.

"Don't you dare do that," she said.

"I won't," I said, feeling a twinge of guilt. I had already made up my mind to go. I had to get out of the Northeast. It was now or never. Besides, I couldn't afford it in New York anymore. Actually, I never could.

Two months later, I loaded my word processor and two suitcases into my banged-up Buick Skylark and headed south on I-95 for a one-month "sabbatical" on Miami Beach, as I told my publisher. I landed at an art deco apartment-hotel on Twenty-First and Collins, the front porch crammed with rows of elderly people on folding lawn chairs. "She's an edituh from New Yawk," one woman whispered loudly as I passed by one morning in my bikini and straw hat, headed to the beach one block away.

I could do this, I thought, gazing at the rolling waves, the cloudless blue sky, the seagulls poking for mollusks in the purple-gray sand. I could get a cheap apartment here and find some kind of nothing part-time job and devote myself to writing. What was the worst that could happen? Even if I failed, it wasn't like anybody here would notice. I'd be just another lost soul wandering the beach in a sarong or scribbling poetry on Espanola Way or throwing down shots at The Deuce. Everyone I met seemed to have washed up on South Beach after some crisis or humiliating failure.

Back in my hotel room, I plopped on the sagging bed and pulled the worn rotary phone toward me. One, two, three rings for good luck. Today was my thirtieth birthday.

"Hello, Cherry Lane Music," the receptionist answered.

She had that thick, taffy-stretched accent that felt as familiar and comforting to me as a warm, soft blanket. But I didn't need to live in New York or anywhere near my father to feel at home now. I had found my tribe of exiled New Yorkers and fellow misfits. I was putting down new roots on a tiny spit of sand curving toward another continent, buffeted by trade winds and hurricanes and the frenetic rhythms of merengue and salsa. Something interesting was bound to happen.

Luciérnagas

February 1993, Miami Beach

Warm salt breeze on my face, I crouched at the base of the weathered door and twirled the rusty combination lock right until it clicked open. Grasping the stiff rope handle, I yanked the heavy wooden panel so it swung toward me forty-five degrees. Ducking underneath the door—peeling aqua paint outside, steel-reinforced wood inside—I propped my palms overhead and pushed upward with all my strength, straining to reach the two metal hooks dangling from the awning above.

My thin arms—now sunburned and freckled after a year and a half of living on South Beach—trembled as I pushed the fifty-pound slab toward the sky. Just one more inch, I prayed, reaching for the hook on the right before my arm buckled. The door plunged to within an inch of my skull before I caught it, sending a searing jolt through my wrist. Fighting despair, I propped both palms under the door and tried again. I was standing tiptoe this time, arms shaking, reaching, reaching; first one hook, *snap*, then the other, and—phew, I had done it. I had opened my beachfront cabana at the Shelbourne Hotel on Collins Avenue and Eighteenth Street.

I stepped inside, tossed the battered lock on the desk, and collapsed onto the pink chenille armchair by the door, like I did every morning. I was at my writing office. I was safe. The irony, however, was not lost on me. I had arranged it so every day I had to risk a concussion to write.

And it was all my fault. Because I had done a stupid thing.

My first month on the beach, I had fallen for a cute but messed-up guy I met in a Cuban diner and impulsively put down all of my savings to rent an apartment with him. Our relationship, if you wanted to call it that, fell apart in a few months, but not before he refused to give me back my first-last-and-security. I was staying on platonically as his roommate until I could scrape up money for my own place, a process that was taking a while. So, I crashed at the apartment at night and worked at a bookstore in the afternoon, and mornings I wrote at the cabana that I rented for sixty bucks a month. Because I couldn't write with him in the room. Correction: I couldn't write with any man in the room. One of the fucked-up neuroses I had been dragging behind me, unseen, since my twenties. Like that seagull out there under the lifeguard's chair, one wing trailing in the sand, limping after its mate.

Which reminded me. I still had the business card in my back pocket. That could wait, though.

I turned on my word processor and slipped in a floppy disk. I was working on short stories again. I had just gotten an acceptance letter from another literary journal. It was exciting and slightly unnerving to see my byline on something besides a magazine article—unnerving because of another embarrassing phobia I did not want to dwell on this morning. To shut off that train of thought, I pointed my cursor at a file marked "Winot" and lost myself in conjuring the world of a Haitian fruit seller on Coral Way as the sun rose higher over the beach, higher and higher in a sky of cerulean blue, while at the far corners of my attention, something else tugged at me.

A fly buzzed through the open doorway and banged against a wall, then into the lamp, before finding its way out again. Oh well. It was no use. I was done for the day.

I fished it out of my pocket and smoothed the bent corners. "Jorge Vera, disaster specialist, American Red Cross," the card read. What an outlandish job title. Should I be offended that someone calling himself a disaster specialist had seen me nervously flitting around a crowded party, in pre–panic attack mode, and singled me out for his attentions? Did I look like a disaster in need of fixing?

A hot breeze blew through the open door, bringing a gust of sand. He had sleepy brown eyes, smooth olive skin, a fine Roman nose. The whitest teeth. Dark feathered hair—a funny white streak in the middle—to his shoulders. Long, muscled legs in shorts—a runner, maybe?

He had said the corniest things: Bet I can guess your sign, I'd like to show you my photographs, When can you come to my place so I can cook you pasta with seafood? He smiled—obviously he knew they were pickup lines—but there was something intriguing about him, a quiet, bemused calm. I kept going from room to room of the large house filled with artists and socialites and drag queens, searching for an exit. He would reappear at my side. The contact was fleeting, like the brush of a butterfly's wing.

"Perhaps you could pose for me sometime," he said. "I've been getting back into photography since I came to Miami after Andrew."

I ignored his question. He smelled like lightning, rain. "Where are you from?" I asked, trying to place his elongated vowels. We were on the second floor, on a balcony overlooking a large backyard. Tiny white lights blinked below, chased by dark figures, their arms raised.

"Lima, Peru. I left because of Shining Path. You know, the terrorist group."

"No, I don't know, sorry," I said, feeling inadequate.

He slowly breathed in and out. One finger tapped on the railing. "You know what those are called in Spanish?" he finally said, pointing to the lawn. "*Luciérnagas.*"

I tried to say it. We both laughed.

"I remember, as a boy, seeing them in the jungle. Their lights were this big," he said, measuring the top joint of his thumb.

"Really?" I tried to imagine it. Glowing thumb-sized blobs floating through the black rainforest, toucans and howler monkeys calling overhead. How surreal.

"The ones where I grew up in New Jersey were tiny and skinny, like this," I said. "But we thought they were magical too."

"Yes, magical—probably still to this day, right? The things that happen to us when we're kids, good or bad, can be so haunting."

You're telling me, I thought bitterly. What time was it, anyway?

"Well, I have to be going," I said out loud. "I have to get up early and write."

"Oh, so you're a writer. Maybe someday you can show me your work. Here's my card," he said, pulling something out of his wallet. A burst of fireflies swarmed around the bushes below. "I'd really like it if you called me sometime, Barbara. *Qué bárbara.*"

I can't get involved, I thought now in the cabana, tracing a finger over the shiny raised letters. *I've had enough of men.* The one passed out back in the apartment bragged he didn't smoke pot and turned out to be a cokehead. The one back in New York was a know-it-all editor who kept pointing out my grammar errors and gave me migraines. I can't. I can't. I can't.

Mangoes

Sunlight flickered on the rumpled sheets, filtered through the jalousie windows, palm fronds clattering in the breeze outside. I was lying on my stomach, Sunday-afternoon sleepy, as he traced a finger down my bare back. We had been together almost a month. I had eaten his seafood pasta. It was good. I had seen his photographs; they were even better. The black-and-white prints were pinned to the walls of his bedroom, and in them, I saw deserted midwestern highways and human shadows cast on wide, barren plazas, and ghostly, billowing curtains in an abandoned button factory.

Loss. Beauty. Things outside the frame.

He fetched something from the kitchen and placed it on the nightstand. I rolled on my side to peek: It was a glass bowl filled with the mangoes we had bought at the wholesale market downtown. Dusky green peels with blush-red spots, deep-orange fruit inside. He pierced one with his Swiss Army knife, releasing the sweet, ripe smell. I flopped on my stomach again. I could hear the blade slicing the fibrous flesh, wet pieces dropping in the bowl.

"Stay there," he said quietly.

He lifted my tangled mane of hair and tucked it under my chin. Gently, he placed pieces of mango on my back. Across my latissimus dorsi, down my spine, one cool slice at a time. The fruit's sticky wetness warmed to my skin, the juice slowly seeping out. Out of the corner of my eye, I saw him pick up his Nikon.

"Can I?" he asked.

"Yes," I said.

Click, click, click. He crossed to adjust the thick louvered windows. From the back, he was sleek, tanned, smooth, otter like. He ran five miles a day. I had been right about that.

"My nose itches," I mumbled into the pillow.

"Try not to move." He handed me a tissue.

The mango's warm stickiness trickled down my back, over the curve of my hips. He discarded the camera. I felt his tongue gliding down my spine, warm, wet, and faintly grainy, like the mango. He kissed the hollow dip at the base of my spine, then flipped me over. Our tongues met. He tasted sweet, tropical, slightly perfumed.

He reached into the bowl and scooped out more pieces. He arranged them in the hollow of my stomach, from hip to hip. I was sighing.

The palm fronds clattered louder by the window, empty husks. A storm was coming. It all felt strangely familiar. I wasn't going to say it out loud. I was home.

Over the next few months, we spent most of our free time together. On the beach. On Lincoln Road. In his apartment. In my new studio apartment. He was attentive, relaxed, a good listener, less anxious and less talky than me. But there was a muted pain. Divorce, two children, a wife in the Midwest who wouldn't let him see them. He retreated to meditate for an hour every morning; I stole off to write.

I spent hours looking at his silver prints. There were his young son and daughter, buried in the sand. A laughing head, a bent arm, a footprint. Scattered elements, nothing whole.

At night, I curled behind his back, grown wide from rowing crew in high school, at a club in the port of Callao. Sometimes I pretended he was a whale I was hiding behind. Safe.

One day, he was looking through an old photo album of mine. He stopped at a snapshot my father had taken at a cousin's wedding in the mid-1980s.

"Why are your hands covering your face?" he asked.

Very carefully—like I was tiptoeing through a minefield—I explained what had happened in my twenties: how my father had become infuriated when I quit the oboe to become a writer, how he had cut off all financial support, his denying all of it years later. I tried to keep my voice steady. I did not want to sound melodramatic.

"So, yes, I wouldn't let him photograph me for a couple of years," I finished.

"Hmmm," said Jorge. "So . . . how many brothers and sisters do you have?"

"None. I'm an only child."

He looked at me searchingly. "So, let me understand: Your father punished you because you wouldn't become a professional oboe player? Is he a musician?"

"No, but he loves classical music. He has, like, three thousand albums."

"I see," he said skeptically.

"He was just bummed out," I hurried on. "He had invested a lot in me becoming a musician. I studied at Tanglewood, in Fontainebleau. You know—"

"No, I don't know." He flipped to a photo of my parents at the wedding buffet table. My mother looked tense and jittery. My father's face was red with drink and exhaustion.

"He drinks," I explained, feeling ashamed. "Not a lot but every day. He worked for the post office for thirty-six years. He's retired now."

"Huh. A retired postal worker who drinks," he said, thoughtfully, like Colombo retracing the steps of a crime. "So, he sold your instrument and cut you off. His only daughter. Because you wanted to be a writer."

"Yeah." My lower lip trembled.

He met my gaze. "You know it's weird, right?"

"Uh-huh," I said, uncertainly.

"Fathers don't do things like that to their daughters. At least not where I'm from. What does he say now that you're published in magazines and newspapers and those literary journals?"

I looked down at the picture of me in the pink silk dress with my face buried in my hands.

"Nothing," I said. "He has never said a single thing about it."

A few weeks later, we were at my new studio, and I went to my computer and Jorge was in the bed and I started writing and that was that.

Transitions

We got married (officially) three years later, in 1996, on Miami Beach, in a beachfront hotel ten blocks north of my old writing cabana. Thirty of Jorge's relatives flew up from Lima, and my parents and a few relatives on my side came, along with our friends. My mother was thrilled to have Jorge as her son-in-law, and my father was polite and distant and increasingly tipsy as the night wound on. When the salsa music cued and Jorge and I stood up to dance, my father's eyes visibly widened; apparently, it was just hitting home that his daughter had married a Latino.

By then, I was supporting myself as a freelance writer, and Jorge was living a double life—Red Cross disaster expert by day, artist-in-residence at the Art Center South Florida by night. Two years later, burned out by disaster work, he quit "the Cross," as he called it, to do photography full-time, and when I was offered a fellowship to study creative writing at the University of Florida, in 2001, he backed me up. We moved five hours north to Gainesville, bought a small house, and adjusted to a quieter way of life, with Jorge accepting a job as the director of an environmental nonprofit.

At my mother's insistence, a year later she and my father sold their home in Toms River and bought a four-bedroom house near ours. It was so big, two families could have fit inside. My mother and I had two years together, with my father still keeping his distance. Then, in June 2004, one month before I received my master's degree, she died of lung cancer, just six weeks after her initial diagnosis.

I was flattened by despair, a kernel of which never left me.

Not long after the funeral, Jorge was offered a better-paying job with a nonprofit in Miami. "We should stay here to look after my father," I said, feeling guilty about having coaxed him to Gainesville, only to abandon him two years later. Jorge reluctantly agreed to stay. My father had been boozing it up since the funeral, sometimes starting before noon, but otherwise, he seemed to be coping, going to church and becoming a higher-up in the local chapter of the Masons. He still wore his suits everywhere, even to the supermarket, in conversation alternating between awkward pauses and non sequiturs. "Sir Francis

Drake had eleven brothers and sisters," he would announce to the drugstore clerk. "He died at sea, you know."

Six months after my mother's death, my father bought a miniature gray-and-white poodle—a nervous thing that peed on the carpet when it was startled—named it Charlie Brown, and the two of them became inseparable. At midnight, they would sit together in the cavernous living room, watching Turner Classic Movies and eating donuts, which my father fed to the dog in generous slices. "I don't eat sweets," he would tell me and Jorge solemnly, the white-green-and-red Krispy Kreme box in plain sight on the microwave. "And I certainly do not give them to Charlie Brown."

Within a year, the dog had to undergo an expensive kidney operation.

My father was cranky and self-sufficient, grateful for Jorge's help with home repairs but bristling whenever I tried to wipe up the dust that was accumulating everywhere. He began covering every inch of table and counter surface with letters and magazines; when I suggested he hire a regular cleaning service, which he could certainly afford on his government pension, he gave it two tries and accused the housekeeper of stealing his "best thimble." He later "found" it in the bedroom but didn't hire her back. I could imagine my mother, somewhere in the afterlife, puffing on an astral cigarette and shaking her head.

She would have had a fit over his meals: lunch at Wendy's, Boston Market takeout for dinner. Once a week, he bought a piece of cod from Publix Supermarket and baked it according to the directions printed on the label. I had to retrieve the grease-caked pan from the back of the cupboard where he hid it, unwashed, behind my mother's dusty muffin tins.

He yelled at me for washing up. He yelled at me for throwing out year-old, unopened containers of Breyers Yogurt. He wasn't keen on having me cook. It struck me he wasn't making good decisions for himself, but I chalked it up to the eccentricities of old age.

Then, in 2006, the funding for Jorge's nonprofit dried up, and he was out of a job. We began a rapid slide into poverty, barely kept alive by my mouse's wages as an adjunct English professor at a community college. There was no going back to magazine writing; the internet had killed that market. Taking out an extra ten thousand dollars, we refinanced the mortgage on our home, a

1950s ranch house twelve blocks from the university stadium. On game days, the air throbbed with ninety-thousand fanatics screaming, "Go Gators!"

Dreamy ads for sleeping pills played nonstop on TV during that time. "Can't sleep? Can't believe you're thinking about sleep? Ask your doctor about Lunesta." Ambien. Restoril. I couldn't ask my doctor. Like many Americans, Jorge and I couldn't afford medical insurance.

Mortgage companies began sending us overnight offers to refinance our house again, this time at 200 percent, 250 percent, of its value. Enclosed were pretend checks whose amounts made my head spin.

Reluctantly, I tore up the offers. It was 2006, edging toward 2007.

Something had to give.

"That Wouldn't Be a Good Idea"

November 2006

"Why won't you ask him?"

"I can't do it," I said, avoiding Jorge's gaze. We were sitting on our back porch, facing the wooded backyard I loved so much—the graceful live oaks, the tall pine trees, the deep-green azalea bushes that became frothy with pink blooms in the early spring, the noisy cardinals and perky sparrows flitting from birdfeeder to nest to birdbath.

"Why not?" he persisted.

Because I promised myself long ago never to move back in with him, even when times were tough, I thought. *I can't let him crow over our failure. Okay, my failure too. As an adjunct, I make less than a Walmart worker, with no benefits.*

"Because I just can't, that's all," I said out loud, looking at him.

"I don't get it. He's your father."

"So? This is the guy who would only give us two thousand dollars for our wedding."

"Yes, but this is different. Aren't fathers supposed to be there when you're in trouble?"

"Are you kidding?" I said. "This isn't Latin America with all that *mi familia* stuff."

Jorge stiffened.

"My father is old-school American," I continued. "You know, the Greatest Generation, self-sufficiency."

"No, I do not know," he said icily and looked away.

Uh-oh, I thought. *It's coming. The cold-shoulder treatment.*

He stood up, stretched his legs, and peered at the sky through the window. "I think I'll go for a bike ride in San Felasco," he said.

Great, I thought. *Go biking while I spend my Sunday grading crappy student essays.*

"How many jobs did you apply to this week?" I asked, feeling small and spiteful inside.

"Nineteen," he snapped and disappeared into the hallway. He stomped to the garage, where he had rigged up a darkroom next to his bike rack. Lately

A pilgrim at Qoyllur Rit'i hoists a large block of ice just carved from Qolqepunku Glacier, in June 2006. Ice-carving rituals were banned by the Brotherhood of Qoyllur Rit'i in 2004, but they continued unofficially for several more years. By 2008, the practice had largely died out. Photo by Jorge Vera.

he was spending hours there, developing prints from a trip to Peru we had taken in June with money left over from the mortgage refi. Our last hurrah as people with interesting lives. A trip to see his brother and sister-in-law, Henry and Mariella, in Lima and to visit an ancient glacier pilgrimage at Mount Ausangate, near Cusco—Qoyllur Rit'i, the festival was called in Quechua. The Shining Snow Star. An Andean mountain spirit had been worshipped there before the time of the Incas. The Lord of Qoyllur Rit'i, who was and wasn't Jesus Christ. The pilgrimage happened each year, in May or June, when the Pleiades reappeared in the night sky. Villagers dressed as half-bear, half men— *ukukus*—communicated with the spirits on the mountain in high-pitched voices. A shaggy-haired anthropologist had told us about it over canapés and wine at Henry's house.

The Volvo started up; I wouldn't see Jorge for hours.

I flicked on the darkroom light in the garage. A fresh print was hanging from the drying line: an Andean man in a sweater, jacket, and knitted Nike cap, hoisting a translucent slab of glacier ice on one shoulder. He stared into the lens, a little warily, a little defiantly; this ice was his birthright, to carry down the mountain, to share with family, friends, his village, even. It was bestowed by his "father," the mountain spirit Apu Ausangate—life-giving water to cure the sick, bless the crops and animals. Priests from the cathedral in Cusco even melted the ice for holy water.

In Andean cosmology, the ice and snow on Mount Ausangate were considered the *apu*'s semen. I wondered what the Catholic priests made of that interpretation.

But recently, the lay Catholic brotherhood that oversaw the pilgrimage had called off some of the Snow Star's ice traditions. No ice race to the bottom of the mountain by the bravest *ukukus*. No bringing the ice to the cathedral. Too dangerous, the *ukukus* had told Paco. *El cambio climático, por supuesto.*

I looked closer at the man in the knitted Nike cap. He had claimed his birthright this year. But how long could his people continue their age-old rituals?

I breathed in deeply. I had fallen in love with a glacier. I hadn't even told Jorge.

Below the drying line was Jorge's computer, open to his email: "We are sorry to inform you the position you applied for Sept. 14 has been filled by another candidate," the message began.

Shit. Another one. Why couldn't he get a job? I knew why. It was the mid-2000s, and we were stuck in North Central Florida, and jobs here were for Gators and people with master's degrees, and Jorge was neither. My man was adrift. I should have told him to take that Miami job when he had the chance. This was my fault.

I cried softly in bed that night, curled behind his back, trying not to wake him. The previous week, we had had money for the electric bill or groceries. Not both.

"Okay, I'll ask him," I said the next morning at breakfast.

The following Thursday, we cooked Thanksgiving dinner at my father's house. Charlie Brown yapped and piddled as we lay plates on the quilted placemats—*her* placemats, embroidered with orange-and-gold harvest leaves—on top of the big blue table from my childhood. Then we settled in to eat. There was no prayer, as usual, but there was Beethoven's glorious Sixth Symphony playing on the stereo system, my father's classical albums displayed along the living room wall. Crowning the mantelpiece was an Olin Mills portrait of himself from 2010, his comb-over freshly dyed, his nose powdered to cover the veins, a gold-and-black Masonic pin gleaming in his jacket lapel. Last Christmas, he had given me the same portrait in a silver frame. When he wasn't looking, I snuck into one of his guest bedrooms and slipped it into a dresser drawer, face down, underneath some unopened packages of dress socks from JCPenney.

When dessert was over, Jorge's knee nudged mine under the table.

I could barely say it. "Dad," I began. He was cutting a Whitman's caramel chocolate into quarters. The cheap steak knife clattered on the ceramic dessert plate.

"Jorge and I have a big favor to ask. Could we move in with you for a bit while we get back on our feet?" Ouch.

He looked up at me, startled, as he popped a chocolate square into his mouth.

"We can't pay our mortgage, and Jorge still can't find a job. We could rent our house in the meantime, and we could help you out here, with the house stuff—take care of the pool, help with the cooking, the housework."

The muscles in his jaw worked as he turned the chocolate square over and over. Charlie Brown leaped up, yapping for a bite. The grandfather clock ticked loudly as we waited.

My father stood up and shuffled to the bedroom. He was wearing a suit and tie and dress shoes, as usual. He emerged with his wallet. Against my better judgement, my heart leaped.

"So, John, is it okay if Barbara and I move in with you for a few months?" Jorge asked. He was tipped back in his chair and looking my father calmly in the face, not crumpled over his half-eaten plate, like I was.

My father's eyes darted furtively at Jorge. "Oh, no. That wouldn't be a good idea."

I felt dizzy. "Please."

He picked up the last piece of chocolate and fed it to Charlie Brown. "Yes, you're a spoiled boy," he crooned to the dog. "You get everything you want, don't you?"

"Here," he said to me, pulling something out of his wallet. "This should tide you over."

He laid it on the orange-and-gold placemat. Two twenty-dollar bills. And a coupon for Boston Market.

Yellow Seals

"Take one last look," the man in the sweaty green T-shirt said.

Jorge and I peered inside at the towering jumble of furniture—our Ikea bed, the Italian marble server, the olive-green sofa, cartons of books and CDs, pots and pans, our antique dishes with the magnolia pattern, Jorge's camera equipment. What exactly were we checking—that nothing was damaged on the Gainesville end? That we really wanted to go through with this?

The moving guy clomped down the ramp in his heavy work boots and disappeared into the truck's cabin.

Jorge put an arm around my shoulder. "Well, this is it."

I swallowed and said nothing.

"He'll be okay," Jorge continued. "He has his friends at his church and his Mason buddies. The doctor said he would email us if there was any problem. The neighbors will check on him. And he has the whatchamacallit."

"Life Alert," I said, picturing the medallion hanging around my father's neck. In the middle was a red plastic button he was supposed to press in an emergency and medics would come to his house. Jorge had helped him set up the base station and recurring monthly payments.

My father would rather pay the fifty bucks a month than have us live with him. I was drained of all anger. It was what it was. I never wanted to live with him anyway. So why did I have this nagging guilt, like I had overlooked something?

The moving guy reappeared with the paperwork and a handful of metal tags with numbered yellow tubes. Jorge scribbled his signature. He had horrible handwriting. That was because he was naturally left-handed, but the nuns in Lima had forced him to switch to the right. They smacked the left-handed kids on the knuckles with a ruler.

Jorge ran up the ramp to wedge a rolled-up rug against his photo enlarger. I had never seen him so energized. He loved change, I was realizing. Me? I was more cautious now that I was forty-six. I was still capable of taking big chances, as this gambit showed, and I was excited to take a stab at environmental reporting, as Jorge and I planned to do, but my subconscious betrayed me. For the last month, I had been grinding my teeth at night, my dreams full of cracking glaciers and unintelligible conversations. I woke up in a sweat, the right side of my jaw still moving on its own. It was peculiar and disturbing.

Concentrate on the glacier, I told myself. *You'll pick up more Spanish down there.*

The moving guy climbed up the ramp and pulled one door shut, securing it with two metal bolts. He slammed the other door shut and fastened more bolts. He slid the tags through the bolts and capped them with the yellow tubes. We were sealed.

"Take a picture," the guy said. "You'll want to check that against the bill of lading."

Jorge photographed the yellow seals, numbers 074953, 074954, 074956.

Brown dust from the driveway kicked up as the huge truck backed out. It was a hot July afternoon, and we had sold our ranch house, used the proceeds to pay off our mortgages and dumped almost everything we owned in this shipping container, with a few thousand dollars remaining to start all over.

The next time we saw those seals again, we would be breaking them open on the docks of Callao, Peru.

Two pilgrims in traditional Andean dress climb the path from Mahuayani to the Qoyllur Rit'i shrine in June 2006. Photo by Jorge Vera.

The *Ukuku's* Wedding

June 2006, Sinakara Valley, our first Qollyur Rit'i pilgrimage

We have been here for half a day, and my senses are overwhelmed. The pounding of the drums, the dancers whirling in their spangled costumes, the hundreds of candles flickering in the ice-cold sanctuary, dusky incense wafting around the kneeling pilgrims, the wheezing flutes and out-of-tune trumpets, the throbbing in my head, my queasy stomach, the drums, the drums . . .

I climb up the mountainside to an emergency tent run by the Peruvian civil defense corps. A Dr. Victor Andía sees me—a calm mestizo man in his thirties or fifties, I can't tell anymore. People are starting to look young and old at the same time. He checks my oxygen levels, says I'm okay, and gives me a pill for altitude sickness. However, it won't kick in for a day, and I have two more to go, which is disheartening.

He pencils in my data on a printed chart. I'm incident number 1,721 in the category "Hurt." Jesus. We're all falling apart here. The other columns are "Drunk," "Dead," and "Births." The last one includes two children, I see, reading upside down. One miscarriage. I can't make out the Dead column.

"Three fatalities," he explains first in Spanish, switching to English. "A Frenchman who got hypothermia, a dancer from Cusco—a rock fell on her head while she was sleeping. An ukuku from Paruro, thirty years old. He was here with his girlfriend and their three children."

"How sad."

"Yes, and maybe no. When an ukuku dies at Qoyllur Rit'i, it is an honor for him and his village. It ensures a good harvest. The priest will have to marry them, though."

I think about this. It sounds like human sacrifice. The "marry" part makes no sense.

His round eyes grow thoughtful. "The local people have been coming to this place for hundreds of years. Since before the Spanish came. This is our way. At some point, the Catholic Church got involved. Now it is both, Andean and Catholic."

He meets my gaze: "Don't think about it too much. It makes sense here."

Two days later, Jorge, Paco, and I climb down the mountain with thousands of pilgrims, all of us crammed on a steep dirt road too narrow for a VW Beetle. People push and shove on all sides, desperate to catch a bus or an open-bed truck back to their province. I'm racing so fast, I trip and nearly fall off the cliffside. That close to becoming another statistic on Dr. Andía's chart.

Rounding a corner, I glimpse a ceremony taking place on the mountainside: A tall, dark-haired priest is standing over the prone body of an ukuku in a green-black-and-yellow robe. Crouched by the dead man's side is a bereaved young woman, with three small children beside her. One little boy is crying. The priest has joined the hands of the woman and the dead ukuku and is muttering some words. Behind them stand a group of eight ukukus, their heads down.

Jorge muscles throughout the crowd to get a closer look.

"He was marrying them," he says when he returns. "Now the ukuku can live in heaven at the side of El Señor de Qoyllur Rit'i."

2

THE DIAGNOSIS

Pigeons

Miraflores, Lima, Peru, January 2011

It was a warm Sunday afternoon in early January—not midwinter, but summer, in Peru—when I began dreaming of blue ice again.

Jorge and I were sprawled on opposite ends of the couch, his long, tanned legs intertwined with mine, our black Lab, Lola, snoring beside us on the tiled floor, as fat gray pigeons—*cuculís*, in Peruvian Spanish—cooed on the windowsill outside. Jorge was fiddling with a new hybrid camera he had bought for freelance work; I was leafing through a *Forbes* I had found in a doctor's office a few days earlier, excited to find something in English to read (with just five months of classes, my Spanish had stalled at an intermediate level). The two-story house we had been renting since 2007 was located in the very neighborhood where Jorge had grown up. Directly across the street, framed by our living room window, was the park where Jorge and Henry had played as children, Parque Leoncio Prado. It was named for a Peruvian mariner who had been martyred in the nineteenth-century War of the Pacific, captured in bed by Chilean soldiers. The towering bronze statue in the center of the park showed Colonel Prado posing jauntily in his battle uniform, sword in hand, not flat on his back in his pajamas.

I peered over my magazine at Jorge, his face knit in concentration as he tested the camera settings. When we met nearly twenty years ago, his hair had been past his shoulders and jet black, save for the white streak down the middle. Now it was cropped short and salt-and-pepper gray, the white tuffet still poking up defiantly. My own hair was long and strawberry blond, a detail that had caused a scene at Lima's Plaza de Acho bullfighting arena two years earlier. As stringers for the *Miami Herald* reporting on Peru's growing anti-bullfighting movement, Jorge and I had been allowed into Acho's fabled *callejón* (alley), the inner circle where matadors prepare to enter the ring. A man in the audience had noticed me standing there, notepad in hand, and, scandalized, yelled to the crowd in Spanish: "A woman in the *callejón!* And a redhead, no less!" I was double bad luck for the *toreros*, as bullfighting tradition had it.

In addition to covering the bullfights, Jorge and I had been to the Snow Star pilgrimage twice since moving to Lima. We had collaborated on a photo story about it for NBC.com, and I wrote about it nonstop on my blog, An American in Lima; that led to freelance gigs "fixing" for *NBC Nightly News, The Today Show,* and *Dateline,* including a special on water wars in the Andes. But despite three and a half years of effort, I had not been able to sell a single article about the glacier or climate change to any publication in the United States, which was still fixated on the so-called debate over whether climate change even existed. Newspapers editors were more interested in stories about cultural change and my experiences as an expat. The "strange" and "exotic" sold back home, even in the literary realm. Before I left for Peru, the Florida Council on Arts and Culture had awarded me a fiction writing grant based on a factual account I had submitted of Mother's Day traditions in the Andes. Perhaps they had thought the drunken mothers carousing on Cusco's Plaza de Armas were figments inspired by magic realism.

The real was not real in the United States in 2011. And when it came to selling environmental stories, I lacked the science writing credentials to persuade editors to give me a chance. So, I just kept blogging and selling articles about "local color," while Jorge turned to art photography, renting a studio in a crumbling colonial-era mansion downtown. For money, he helped produce programs on the environment for European TV stations, whose audiences did take these things seriously.

He was lucky to have those connections, I thought, enviously. All I had was an offer to teach English part-time at the Peruvian University of Applied Sciences (UPC), in the district of Monterrico. Hah, like I wanted to do that: endure a forty-five-minute cab ride each way to teach verb tenses to students at a for-profit university plunked down in the dirt mountains by the house of Jorge's elderly uncle and aunt. "No, thank you," I was planning on telling the university.

Jorge angled the camera lens toward my bare foot, which I shoved self-consciously under a cushion. My left big toe was sporting a thick yellow callous where the hiking boot had rubbed.

"Hey," I warned.

"Don't worry, I'll delete it. I'm just trying out the exposure settings."

I glanced at the essay I had been trying to read: "The Big Climate Hoax."

"*Forbes* just published another stupid piece," I began, citing the reporter's name.

"Oh, that *huevón* [idiot]," said Jorge.

"Do you want to hear the lede? It's all lies. This bullshit needs to be called out."

A tiny spasm of annoyance creased his cheek. "I thought you were done with that."

"I'd still like to sell a story on Qoyllur Rit'i. Climate change might not be happening yet in the United States, but it soon will be. Lonnie Thompson told me so at Huaraz."

He shook his head. "Look, Barbara, we went up to the glacier three times with Paco. I had my photo show downtown; we did that photo story together; you did that thing with Anne Thompson. We've done our bit for the environment."

"But my story—" A flurry of anxiety rose in my chest.

"Stop it with that face. Hey, did you see this?" He flashed the camera screen at me. "The glacier came back."

It took me a second to recognize what the bulbous white-and-yellow mound was: my toe.

"Jerk," I said, throwing a pillow at his head. It missed and bounced next to Lola, who leaped up, tail wagging.

"Hey," he said. "Let's go outside so I can try out the camera."

The sun warmed our faces as Jorge, Lola, and I crossed the lane and entered the block-long park. The bisecting pathways were crowded with families enjoying a stroll after their 1:00 p.m. main meal (*almuerzo*): children skipping alongside their parents, nannies pushing strollers, grandparents held by the arm by young women in white scrubs—home health aides known as *enfermeras técnicas*. The women walked at a snail's pace alongside their elderly charges, their white shoes measuring the length of each polished concrete paver.

Scrawny gray squirrels chattered in the trees overhead as we joined the Sunday afternoon procession. Jorge's shutter fired as he surreptitiously took a flurry of photos over one shoulder, not even looking through the lens. Lola sniffed at the grass as we traced our usual route along the walkways: down to the yellow D'Onofrio ice cream cart, past the big house once owned by the family of a childhood friend who used to invite him after school to eat vanilla wafers, along the dusty shrubs where Jorge and Henry made a fort one summer and waged a month-long war with the neighborhood kids, back to the center of the park and to the heroic Colonel Prado, his bronze epaulets dabbed with pigeon shit.

A scuffed soccer ball bounced in front of us; Jorge stuck out a foot and kicked it back on the grass.

"Gracias," a teenager called.

"De nada."

"Let's stop here," I said, nodding at shaded bench.

I leaned back against the cool wooden slats, *cuculís* cooing and humping in the branches above, as Jorge discreetly filmed the people drifting by. Our German neighbors waved at us from a nearby sandpit, where their youngest child was digging with a spoon. In the flower bed next to them, a sweet-faced health aide was coaxing an older man out of the marigolds.

"Come here, señor," she said in Spanish. "This way." He shook his head. "Please, señor. We can have ice cream."

I would never have the patience, I said to myself. Never.

I thought of my father, anger and guilt rising. No, my father did not want our help. Nor did he want to help us. Anyway, we had done all right here on our own, no thanks to him. That was how he wanted it: separate households, separate lives, an annual visit in which we flew up to Florida for the holidays— stilted interactions, him in his bathrobe half the day, playing solitaire at the kitchen table—at the end of which he looked visibly relieved when we pulled our rental car out of the driveway. His life was an endless round of doctor's visits and Masonic meetings, trips to the vet, evenings with Charlie Brown and Turner Classic Movies. Who needed a daughter or a son-in-law when you were equipped with such plentitude and Boston Market takeout?

I could still picture the coupon tossed on the woven placemat. The orange-and-gold placemats of my beautiful dead mother, who was never coming back. Hole in my heart.

That had been back in 2006. The first year I saw the melting glacier and its turquoise-blue heart up close. The first time I tasted its ice and let it drip down my throat and let its dizzying chill lodge in my chest, a sensation that had never left me, if I was honest. Years later, I was still obsessed with a fifty thousand–year–old glacier. A feeling he didn't share. We'd fought about it on our last pilgrimage, in 2009. Him and me, screaming in the tent. Paco got it, though. Our nervous, stubborn guide who herded llamas and alpacas in the shadow of Mount Ausangate.

"I want to go back," I said out loud.

"What?" Jorge frowned as he looked away from his viewfinder.

"To Qoyllur Rit'i. I want to help out Paco and his family. I think your cousin Chata might want to come too."

His face softened. "Barbara," he said. It was a new face. *Pitying* was the word that came to mind. "Let it rest. You have that university that wants you to teach. Why not accept?"

"But I can get an assignment . . . 2011 is supposed to be my year. My horoscope said—"

A muscle twitched in his cheek. "You see this camera? I am going to use it for actual paying gigs, like the one with ARTE next week. I am not blowing eighteen hundred dollars on an expedition to some glacier in buttfuck Quispicanchi that nobody in their right mind cares about, besides Paco and his annoying mother."

"I care about it."

"Great, you pay for it."

"I will!" I stood up, bumping my head on a branch. A startled *cuculí* flew out and landed on the sidewalk, fluffing its feathers. I grabbed Lola's leash. "You'll be dying to come along when you see the kind of interest I can stir up with this, this thing I'm going to write."

— ❖ —

The dining room was empty and still, shadows shifting in the high ceiling overhead. I unhooked Lola and let her flop down on the cool tile. Why did I get so worked up over these things? It was only a story, a story I probably wouldn't sell. You win some, you lose some. So why couldn't I give this story up? Why did I insist on fighting about it with Jorge, of all people, the person who had convinced me to climb the glacier in the first place?

I walked over to the Italian server and rested my palms on its cool, marble slab. "Cálmate," it seemed to be saying, or whatever the Italian was for "chill out." Observing me from the top shelf were the ancestors, as Jorge and I called them, photographs of my mother and Jorge's parents (long dead), plus plaster statues of San Hilarion, the money saint, and San Martín de Porres, the first Black saint of the Americas. I picked up San Martín and stroked his nubby head; it was gritty under my thumb, the painted surface embedded with sand. He was the patron saint of barbers, sailors, and lost causes. Was that what my Qoyllur Rit'i struggles had become?

I set San Martín back on the shelf, beneath a photograph of pilgrims posing in front of the glacier during our second trip to Qoyllur Rit'i, in 2008. In just two years, the glacier had fled another forty feet up the mountain. One cold morning, Jorge and I climbed to the place where I had stood at the glacier's edge, two years earlier, and found nothing but dirt and moraine. All the ice was gone—the huge, frozen, whale-like wall I had leaned against was no longer there. It was the strangest feeling, to stand where an enormous glacier had been and now was not. The effect was sudden, sharp, bewildering to the body—like entering a room expecting to see someone you love only to remember they're dead. *So, this is what the effects of rapid climate change feel like up close,* I had thought, staring dumbly at the raw brown dirt, my body filled with a disorienting sadness I could not name. I was standing at a place that existed but no longer existed. I was there, but there was no "there" there.

Could you be homesick for a place that was vanishing into thin air? I wondered, tracing a finger over the plexiglass.

An Australian philosopher had coined a word for this. *Solastalgia.* The grief induced by seeing a familiar place gradually destroyed by climate change. It felt so personal, a crushing sadness mixed with helplessness and bewilderment. Standing where the glacier wasn't had aroused a sharp, almost violent longing in me that day. *Maybe if I close my eyes and think hard enough, this nightmare will end—the ice will creep back down the mountain.*

Five pilgrims in feathered Amazonian headdresses were standing shoulder to shoulder in the harsh sunlight, the shrunken glacier behind them. *Chunchos,* their ceremonial role was called. The sad-faced older guy on the left had suggested they line up where the glacier used to end. He'd been coming to Qoyllur Rit'i for twenty years. Not even his teenage son was smiling. What was there to smile about? "This photo will appear in the American magazines, yes?" the sad-faced guy asked me afterward. "So the *americanos* will understand and help us?" Yes, yes, I had lied, telling myself I wasn't lying.

I had let them down in 2008 and again in 2009. Was I about to do it again?

The doorbell buzzed, jolting me out of my brooding. Waving at me through the front window was a tall, thin woman with chestnut-brown hair and a tranquil smile. It was Jorge's older cousin Chata, who lived with her husband on a horse farm outside Lima. In the early 1980s, she had studied anthropology at Universidad San Marcos, Peru's oldest university. But Shining Path, Sendero Luminoso, had stopped all that. Doing fieldwork in the highlands became too dangerous.

I let her in, raising my head to kiss her cheek: At over six feet, she was unusually tall for a woman in Peru and had played for the national volleyball team. Her nickname, La Chata, translated as "Shorty."

"Cousin," she said in Spanish, handing me a large Tupperware container. "You left this at my parents' house."

I thanked her, blushing; last Sunday, Jorge and I had gone to a family *almuerzo,* and I brought my version of a Peruvian classic, *papa a la huancaína* (potatoes, Huancayo style). It was my first time using fresh ají peppers, and I had miscalculated the spiciness. An older relative had succumbed to a dramatic coughing fit and had to be carted off to the bedroom.

"You came all this way just to give me the *Taper?*" I asked.

"No, Dickie and I are just passing by," she said, pointing to a Mercedes at the curb. "We're on our way to Gorda's." (Gorda, or "Chubby," was her skinny little sister.)

"Ah, good," I said, setting the container on the sideboard.

Chata glanced at Jorge's picture. "I would love to go to Qoyllur Rit'i some-day. It would be fascinating to see the rituals, the pilgrims, the dancing, to observe. Are you going this year?"

"Yes, I think so."

"Well, maybe—"

The car honked outside.

"I have to go. Another time," she said, kissing me goodbye.

I closed the front door and glanced at my watch: 4:05 p.m. Shit. More than an hour late.

I ducked into my office at the back of the house and dialed the Gainesville number, stroking Lola's head as I gazed absently out the window. At the far end of the long, walled backyard was a fig tree supposedly grown from fruit brought from Spain by Pizarro and planted in the courtyard of the palace of government in 1536. Tio Jorge had picked some of the figs when he worked as minister of industry for President Bermúdez, in the 1970s, and given them to families in this neighborhood to plant. The tree was stumpy and not much to look at, and the figs were tough, but it was tantalizing to imagine that here in our dusty, pigeon-filled garden was a living connection to the founding of the City of Kings.

Six rings, eight rings. Where could he be? Our weekly ritual. He filled me in on his appointments, on and on about him, Charlie Brown barking in the background. It was okay he didn't ask about me. All he needed to know was Jorge and I were in the "Land of the Incas," as he bragged to his disinterested neighbors.

Twelve rings. Lola nudged my leg with her wet snout. "Where is he, girl?" I asked, opening the back door. I peered up at the twenty-foot-high walls sur-rounding the yard; cemented to the top were shards of thick green glass that poked out like angry teeth. They had been put there in the 1980s or '90s to deter thieves and terrorists, a common practice in Lima during Shining Path times. There hadn't been a bombing in the city since 2002, but nobody both-ered removing the glass. Limeños just stopped seeing it, like they did the dust everywhere. Maybe if I lived here a few more years, I would stop noticing these things too.

I gave up at eighteen rings.

"Don't worry, he's probably out painting the fence or something," Jorge said when he came inside an hour later. The earlier tension between us had evaporated, as it usually did.

"I'm not worried," I said, lighting a votive on the sideboard. "I'm concerned. How do you say that in Spanish?"

"*Preocupada.*"

"Preoccupied."

"No, worried."

"So, what is *concerned?*"

"There is no difference."

"Yes, there is," I said, looking at the five guys lined up where the glacier wasn't. "Those are two very different words."

Failure to Thrive

Two days later, my father answered the phone. In jumbled sentences, he explained he had been sitting in the lounge chair in his living room "for days." "Something is wrong in my underpants," he said in an anguished voice. Charlie Brown was screaming in the background.

I called his doctor, who contacted the Gainesville EMT and instructed them to go his house. "Failure to thrive" was their professional diagnosis. They rushed my father to a local hospital. A nurse relayed that he was dehydrated and had some kind of infection, maybe of the urinary tract. They put him on an IV and started pumping him with antibiotics.

"I knew something was wrong," I told Jorge. "I knew it, I knew it."

The people who had dreamed up Life Alert forgot one thing. An emergency button dangling from someone's neck is useless if the person in trouble is so out of it he can't even recognize he's having an emergency.

Pinch Hitter

Early February 2011

I looked into his clear brown eyes. "You don't have to do this," I said.

"Yes, I do. One of us has to."

"I could . . ."

"Yes, but your university gig starts in a few weeks," he said. "We could use the extra money. Plus, you don't really want to go, do you?"

I looked away, sobered. He was right. I was a bad person. He loved me anyway.

"Call me as soon as you arrive," I said.

The cab was waiting at the curb. I wrapped my arms around his broad back. Everything was falling apart, but I still had him to count on.

As the cab disappeared, my body relaxed. No, I didn't want to see him. I didn't want to be anywhere near him. The person with the four bedrooms and the vaulted ceiling and the ginormous pool and lanai who said no when we were on our hands and knees.

That night, Chata called, asking for Jorge. I explained as best as I could what had happened with my father. At the end, she asked: "Are you going to Qoyllur Rit'i this year, cousin? Maybe I could go with you?" The question hung in the silence.

Quizás, I said. Maybe.

I hung up the phone. My mind was made up, I realized. I was going to Qoyllur Rit'i.

Two days later, while I was washing dishes, a Cusco number lit up on my phone.

"Señora Vera!" a raspy voice boomed. It was Paco, our guide. Since that first trip to Qollyur Rit'i in 2006, when he had been working as a porter for a Cusco tour operator, Jorge and I had contracted directly with Paco twice to lead our expeditions. To make this call, he had probably traveled to Cusco all the way from his farm, hours away. There were no cell phone towers where he lived, in the village of Upis, 14,400 feet above sea level, beneath mighty Mount Ausangate.

"Do you come to Qoyllur Rit'i this year?" he asked in Spanish. "And Señor Jorge too?

"Yes, Me. Maybe Señor Jorge." The line was all crackly. "What is the date of Qoyllur Rit'i this year?" I yelled. I could never get this part of the festival straight—the dates changed each year, sometimes May, sometimes June, timed with the reappearance of the Pleiades. If the stars could be seen clearly, the farmers planted their crops earlier.

"The *something something* of June, señora."

"What?"

"Yes, yes, señora, we will talk in March. You send the money. We will make Qoyllur Rit'i together." The line went dead.

I thought about Paco and his wife, Blanca, and their four children in their stone hut in Upis, on the treeless puna not far from Qolqepunku Glacier. There were no hospitals in that barren landscape, no stores, no police, no anything. Just a bunch of llamas and alpacas and people growing potatoes in the red-dish dirt. With the money he made guiding people like me and Jorge to the pilgrimage, he could support his family for half a year. With some strawberry jam and a newly slaughtered alpaca, he could have a wedding feast for one of his daughters.

I had wanted to say something to Jorge the other day in the park, right be-fore I bumped my head on the branch. It was about Paco's mother, a shifty-faced woman with two teeth in her mouth. At the end of our last pilgrimage, Jorge and I had returned with Paco to Upis to drop off the leftover supplies at his hut. A dusty burro was tied to a pole wedged in the stone wall, and Paco's mother was waiting for us outside, her stocky body barring the door. She looked pissed.

She dug through his duffel bags, removed the jars of jam and instant soup packets and bags of white rice, and lined them up on the wall to count them. Then she yelled at Paco in Quechua, and he yelled back, wincing at whatever it was she was saying. He turned around and yelled at Blanca, who looked petrified. In the end, he loaded about three quarters of the food we had given him—along with a bottle of Bayer aspirin his mother made him beg us for—into woven pouches on her burro, and his mother hobbled down the dirt road with her haul.

Paco stared after her, a beaten look on his face.

That was what I had wanted to tell Jorge. I had to go to Qoyllur Rit'i this year. Not just for me. Not just for the glacier. But for Paco. To help him outsmart that pushy mother of his.

The Peruvian Reset

"It's bad," Jorge said over the phone. My father was in rehab now. He had been there for two weeks, sharing a room with an angry Russian man who couldn't walk anymore and took it out on the nurses. "The head psychologist says your father has . . . Alzheimer's."

"Alzheimer's? How do they know?"

"They did tests."

"I can't believe it," I said, dumbstruck, but as I was saying it, I could believe it. All those bizarre things he had been doing for years, stockpiling vitamins, hoarding expired yogurts, refusing to clean, firing the Mini Maids over a supposedly stolen thimble, maybe even refusing to let me and Jorge move in with him several years ago—those were not normal behaviors, even for someone as strange as my father. I should have recognized the signs sooner. I should have realized a chronic alcoholic in his late eighties was a prime candidate for dementia. But what did I know about this disease? It wasn't something that ran in our family, or at least that I'd heard of.

"How bad is his Alzheimer's?" I asked. "Is it permanent?"

"'Mild to moderate,' the psychologist says. Whatever that means. Yeah, he'll probably have it the rest of his life—although she said sometimes seniors get dementia after a urinary tract infection and, when that's over, their brains recover."

"The psychologist said the Alzheimer's could reverse itself?"

"Yes, but she said we shouldn't get our hopes up. Oh, and one more thing," he added. "The doctors say he can't live on his own anymore. He will need full-time care. We have to get him a nurse or put him in an Alzheimer's unit or something."

Math.

Father's monthly pension from American Postal Workers Union: less than $3,000.

Full-time dementia caregiver in Florida: $20 to $25 an hour.

Shared room in a dementia care ward: About $4,500 to $6,700 a month.

Money Jorge and I had to cover difference between pension and dementia
care: nada.

Chances of me and Jorge getting decent jobs in North Florida in the recession:
zilch.

We would have to sell my father's house to pay for it all. That was what
families did in America, right?

We went round and round in our nightly conversations, debating the non-
possibilities. We had never been so snippy and irritated with each other. I was
aware of a new feeling toward Jorge: guilt. My problem father had now become
his problem as well. He wasn't complaining, but I knew he hadn't signed up for
this. How would I feel if I were in his place?

I imagined us at our wedding in 1996, standing on the deck of the Miami
Beach Ocean Resort, the freelance minister lady reading from her laminated
sheet of vows.

"Do you, Jorge Alberto Vera Du Bois, take Barbara Ruth Drake and her
demented father, John Drake, to have and to hold, from this day forever, till
death do you part or until you yourself become broke and demented?"

Nobody would ever marry anybody if they knew what was really going to
happen.

Valentine's Day

"It's simple," he said. "We'll move your father to Peru to live with us."

"Are you . . . sure?"

"Of course. That's what you do with elders."

It was like a switch had been flipped in Jorge's brain. The Peruvian reset.
I didn't have that switch, at least not for this parent. I would have taken my
mother in a heartbeat.

Okay, I said to Jorge, grateful, wary, a bit in awe. This was the father-in-law who had kicked us out on the curb, but apparently, the Peruvian reset overrode all that.

Could I be reset too? Maybe all of us could, including my father.

The solution to my father's problem had been there all along, right across the street in the park where the elders perambulated in the afternoon with their helpmates in white. An *enfermera técnica*. A home health aide.

The next week, Jorge and I began to talk about hiring one.

"I asked Henry," Jorge said. "We can get a licensed aide with great references for less than seven hundred dollars a month. You can get two aides to trade shifts. They'll give him great care. They'll give him showers and wipe his butt and treat him like a king."

"Couldn't we just start with one aide? Maybe he won't need so much care."

"Sure. Call Henry or Chata for the details. Have the person start as soon as possible to help you redo the back office. We'll have to use that as your father's room."

"Oh," I said, dumbstruck. "Where am I going to write?"

"Get the gardener to move your desk up to our bedroom."

"Okay . . ."

"And get an *enfermera técnica* fast."

"I'll try. But I can't rush the interviews. You know how iffy my Spanish is."

"You can do it, Barbara. I'm counting on you. You'll see. Your father will like Peru and the food and the pampering, and—who knows? Maybe he'll stop being such an asshole. We'll all be one big happy family. Hah!"

One Big Happy Family, I thought later as I lay on the couch, staring up at the cracked floral medallion on the plaster ceiling. As corny as the phrase sounded, maybe we could achieve our own version of Happy Family. A health aide could keep him out of our hair all week, and I'd be free to write and teach on the side: four mornings a week, UPC was offering me. Jorge could keep doing his production work and teaching. On Sunday afternoons, we could parade my father around the park with the other elders.

I pictured a young woman in white holding him by the arm, gently guiding him down the polished pavers, Jorge and I strolling behind. The anemic-looking squirrels were scampering in the trees overhead, and the *cuculís* were cooing, not pooping, and a less-neurotic Charlie Brown was bringing up the rear with his new friend, Lola. My father's comb-over was neatly plastered over his skull, and he was dressed in a V-neck sweater-vest, like a proper Limeño gentleman.

The only thing I could not picture was the expression on my father's face. Was he happy? Grateful that his long-maligned daughter had graciously taken him in? Contrite even? Or were his thin lips curved downward, the angry Seine?

The Arrival

March 20, 2011

It was close to midnight, and I was standing in the crowded arrivals hall at Lima's Jorge Chávez Airport, my stomach in knots, waiting to greet Jorge and my father and Charlie Brown after their eight-hour flight from Orlando.

With me was our van driver, Gustavo, and Señora Lucinda, my father's new health aide, whom I had found through an agency. Señora Lucinda was a stout, round-faced woman from the highlands in her early fifties, with a kind smile and soft, wavy hair pinned up in a bun. It had been nerve-racking hiring her all on my own, but apparently, I had lucked out. On paper, she was strictly a licensed *enfermera técnica*, but, as I had discovered in the ten days we spent puttering around the house together, painting and outfitting the back room, she had other talents as well. She was gifted with plants and living things, coaxing green buds from the bone-dry herb garden on the patio, and she made the most delicious *ají de gallina* (spicy creamed chicken) I had ever tasted. She was thoughtful, conscientious, and, above all, anxious to know more about her new patient from the United States.

A few days ago, she had shyly asked what my father's personality was like, and I didn't know how to respond. How did you say "evasive and prickly" in Spanish? Did that description even fit my father after his grueling two-month stint in the hospital and rehab?

"He is very sick, but he will be happy to meet you," I said to her.

For tonight's midnight reception, Señora Lucinda had insisted on wearing her full uniform—white smock and pants, white shoes, white sweater—to make a good impression on my father. She had already asked me several times in her low, gruff voice if she looked "professional."

Frankly, I had more to worry about than the impression made by Señora Lucinda's nurse's getup. Two weeks ago, egged on by his Russian roommate— "John, you don't need to make exercises, you are free!"—my father had cut the electronic ID tag off his wrist and slipped out of the Gainesville rehab center. Wearing slippers, shirt, and pants, but no belt, he had wandered down the busy six-lane road that led in the direction of his house, more than two miles away.

By chance, a police officer found him shuffling down the sidewalk, holding up his loose trousers with one hand. Remarkably, despite wandering for hours in the warm sun, my father could remember his home address. Jorge, who was fixing up the house to rent, opened the front door to find my father standing there, sweaty and cowed, held at the elbow by a tall officer.

"I want to go home," insisted my father. "Tell the officer this is my home."

Jorge calmed my father down and convinced him to return to rehab—where, it turns out, none of the staff had even noticed he was missing. Negligence reports were filed with the police, but nothing came of them. We spent the last week of Jorge's stay in Florida wound up like grenades, worrying that my father would make a break for it again, aware that maybe the next time my father wouldn't be so lucky, that he'd go roaming onto the interstate and be lost forever.

A message buzzed on my phone: "Through immigrations. On our way."

I peered over the sea of heads, families waving hand-printed signs in Spanish—*Welcome, Little Manuel!* and *Darling Father, I missed you so much.* Then I saw them: Jorge, in black T-shirt and jeans, a messenger bag slung across his chest, striding along the fenced-in walkway as he pushed my father in a wheelchair with one hand, his other arm lugging a cart of suitcases. My father was dressed in a navy-blue windbreaker, and his mottled pink head bobbed up and down as they dodged the spontaneous family reunions erupting on all sides—hugs, kisses, screams of emotion—at which point I realized, with a shock, that for the first time since I was a girl, my father's elaborate comb-over was gone: someone had finally given him a proper haircut.

Minus the carefully lacquered strands of hair camouflaging his bald skull, he looked all of his eighty-six years: wizened, frail, but surprisingly serene—like an ancient, deposed leader returning to his homeland after a half-century hiatus.

I ran up to them and landed a kiss on Jorge's stubbled cheek. He radiated heat, determination, and exhaustion. It was then I noticed. "Where's Charlie Brown?" I whispered.

"Too much to manage," he said. "I gave him to a lady from your dad's church."

"Does he know?"

Jorge shook his head.

My father's skinny neck swiveled as he peered up at me. It took a few seconds for his watery gaze to sharpen.

"Oh, hello, Barbara," he smiled. His voice held just the faintest note of surprise, as if we were accidentally bumping into each other at the supermarket by his house, rather than being reunited in a third-world airport three thousand miles away.

"Welcome to Peru," I said.

"Peru?" my father said.

Uh oh, I thought. "Dad, this is your nurse, Señora Lucinda," I said, bringing her forward. "She'll be taking care of you."

Señora Lucinda stiffly extended her hand, her fingers glued together like a Barbie doll; my father stared for a few seconds before grasping it.

"Hello, Mister John," she said, haltingly repeating the words in English she had been practicing all week.

"Very nice to meet you," my father said, his automatic manners kicking in.

A smile lit up her round face: *First test passed,* she probably was thinking. Oh, not to know who my father was, to imagine building a cordial relationship with the man.

We wheeled my father to the exit, where Gustavo was loading the van. I pulled Jorge aside. "How was the flight?" I asked.

"You wouldn't believe," said Jorge. "On the Miami-Panama leg, they sat your dad in first class; I was three rows behind in economy. I should have made them put us together. Your father kept getting up during the flight and saying, 'I want to go home.' I had to keep going through the curtain to bring him back to his seat. The person sitting next to him in first class—this prissy Chilean guy in a linen suit—looked really pissed."

He took a deep breath. "Then, in Panama, we almost missed our connecting flight. I had to change your dad's diaper in an airport bathroom. There was no room. It was—"

"—awful," I finished.

I knew those dirty, cramped bathrooms in Panama. The stalls weren't wide enough for a Swiffer, let alone two grown men. I couldn't even begin to imagine the logistics.

I poked my head in the van to check on my father, now installed behind the driver's seat. He was gazing peacefully out the window at the Lima night sky as banged-up taxis and tourism vans wove pell-mell in and out of the crowded drop-off lanes.

"He seems really calm," I reported back. "I would have thought he'd be freaking out."

Jorge patted his messenger bag. "Valium. The head nurse at the rehab center arranged it."

Beside us on the curb, a chubby cop was yelling, "Get going," at Gustavo, who placidly continued helping Señora Lucinda into the van. She squeezed into the last row, leaving the seat beside my father empty. *Of course,* I realized. She was expecting me, his daughter, to sit there.

I slid into the empty space, Jorge ahead in the passenger seat. My father was rolling the hem of his windbreaker, over and over, between his fingers. The doors slammed shut.

Gustavo exited the airport, heading southeast to Lima and the district of Miraflores, about ten miles away. The bright lights and billboards advertising Machu Picchu soon gave way to rows of warehouses as we passed through the port town of Callao, and I tried to imagine what sense my father was making of this unfamiliar landscape. Was he overwhelmed by the noise, the chaotic surges of cars and people at every intersection? Did he notice the layer of dust covering everything, like every first-time visitor to this desert country seemed to comment on? Did he even realize he was in another country? Probably not.

His hand lay limply next to me on the seat. It looked frailer than when I had last seen him. His nails were yellow and ragged. I had never seen them like that.

I lifted his hand and squeezed it. He gave a weak squeeze back. He didn't let go. The tips of his fingers were ice-cold.

"We're glad you've come to live with us, Dad." It sounded like something a normal person would say.

"Thank you," he said in a faint voice. Then, after a longer pause: "It's good to be here."

— ✦ —

We sped down Avenida Faucett and crossed over the Río Rímac, the once-sacred river of the Incas that brings water (and toxic mining chemicals) from the Andes mountains to Lima. Two blocks later, we passed the towering *Virgen del Carmen de la Legua*, a sixty-five-foot-high statue of the Virgin of Carmel balancing the Christ child in her outstretched hand. She looked blandly down on us, her wide concrete arms raised in benediction, as Gustavo swerved to avoid a renegade *combi* bus stuffed with passengers.

Dear Giant Concrete Virgin and Baby Jesus, I thought, sending up an impromptu prayer: *Let this man not ruin my life again. Let him stay safe in his room at the back of the house, and may Señora Lucinda keep him calm and well fed and out of my hair. And let me get an assignment to go to Qoyllur Rit'i so I can finally publish a story lots of people will read.*

We turned onto Avenida La Marina, a commercial zone lined with newish shopping malls and fast-food restaurants and older slot machine parlors and casinos, the latter part of the pro-gaming legacy of ex-president Alberto Fujimori in the 1990s. The neon signs whizzed by in a blur of primary reds, yellows, and blues: KFC, Burger King, Bembos Burgers, Joker's Casino, Texas Station Casino, Palacio Real Tiahuanaco, Hello Hollywood. This was my father's introduction to the real Lima: modern, brash, greedy, growing—one foot in its pre-Columbian past, the other in commercialized American culture.

The tangy sea air stung our nostrils through Gustavo's open window as we turned east onto the Costa Verde, the Pacific coast highway that leads south to Miraflores and Chorrillos.

Somewhere out there in the darkness and crashing waves was the looming bulk of Isla San Lorenzo, Peru's biggest island. In 1579, British privateer Sir Francis Drake launched a raid from there on the Spanish galleon *Nuestra Señora de la Concepción*, nicknamed *Cacafuego*, or "Fire-shitter," by the British. The ship was bearing the annual quota of gold and silver for the Spanish crown, worth about eighteen million dollars today. Drake blew up most of the Spanish fleet, pissed off the viceroy of Peru, and plundered *Cacafuego*, bringing in the biggest haul of his career.

That was the man whose brother, my father loved to brag, was our long-ago ancestor. *El Draque*, the Dragon, as Sir Francis was known in Peru. Now, more than four hundred years later, another Dragon had pulled into port.

I peered at the man slumped next to me, fingers clutching mine, his eyes red and blinking with fatigue.

Was this experiment going to work? Or would it blow up in our faces, like the fleet of the *Cacafuego?* I had no way of knowing.

I released his cold fingers as we sped into darkness.

Pilgrims play on Qolqepunku Glacier in June 2006. The annual activity is sacred to many indigenous people in the Andes, notes scholar Alfonsina Barrionuevo: "The [pilgrims] want to go up to the top of the glacier to pay their respects and play with their Father Apu in the snow. . . . They conduct many ritual ceremonies throughout the year, but they consider Qoyllur Rit'i, at Qolqepunku Glacier, to be fundamental to their worship." Photo by Jorge Vera.

The Author from Cusco

January 2, 2007, Gainesville, Florida

I am sitting at the desk in my yellow book-lined office, looking out at the pine trees and bright-green ferns carpeting the front yard, trying to make sense of what Jorge and I went through at our first Qoyllur Rit'i festival last year. Before me are stacks of books on Andean culture borrowed from the Latin American Collection at the University of Florida Smathers Libraries. I have been slowly combing through the volumes, making notes, feeling myself pulled deeper and deeper into Andean cosmology, still at a loss to understand how all this might connect to the melting glacier and other climate change effects we witnessed at the pilgrimage six months ago.

Inside one book cover is an author photo of a striking dark-haired woman— Alfonsina Barrionuevo, a Peruvian author who has spent her life reclaiming and sharing indigenous knowledge lost to colonization. The fact that she is from Cusco is significant: the area was the cradle of the Tupac Amaru rebellion against the Spanish in the 1780s, and from the 1900s to today, Cusco has been a center of pro-indigenous forces, political and cultural. I dial her number, as we arranged, and four rings later, a lilting, penetrating voice answers.

Is it true she believes Qollyur Rit'i won't survive climate change, I ask?

"When the ice goes, the indigenous people of the Southern Andes of Peru will no longer be able to conduct spiritual cleansing rituals as they have for hundreds or thousands of years," *she answers in English.* "Without the ice and snow, the rituals are not possible. The pilgrimage is doomed to end."

City people, whether in Peru or abroad, do not understand the full impact of what is happening with the glaciers and the earth and global warming, she stresses.

"The farmers and alpaca herders of the Andes, people who work and walk the land, are the ones who talk and know," she says. "They are the ones who think about these subjects. The people from the city are not interested and are totally ignorant about the force, the energy, that radiates from nature toward man, the healing power of the nature-heart, which is only known by the Andean priests. If you ask a city person what an apu is, what the power of the killa [moon] is, he is going to look at you as if you are a Martian because they don't have the least minimal idea. However, if you ask the people of the highlands, they know."

We talk about the ice-cutting rituals at the glacier and how they are now banned—officially since 2004.

"This is tragic for all the farmers and communities since, just as the ukukus do, they believe themselves to be children of the glacier," she says. "They want to go up to the top of the glacier to pay their respects and play with their Father Apu in the snow. These ceremonies are key to the lives of over twenty thousand people who live and work the land in the Ausangate area. They conduct many ritual ceremonies throughout the year, but they consider Qoyllur Rit'i, at Qolqepunku Glacier, to be fundamental to their worship."

"Why is this ice so important?" I ask. As I speak, I recall the icy-warm pang that bloomed in my chest when Paco fed me the glacier ice. It is such a strange and private feeling, I have mentioned it to no one, not even Jorge. I wonder whether to say something about my lingering memory of the ice to Señora Barrionuevo, then stop myself.

"The ice that forms at this altitude is the purest manifestation of the gifts from Pachamama," she is saying, "And this ritual is possible only at Qolqepunku, during Qollyur Rit'i."

"So, what will happen when the snows disappear on this mountain?" I ask.

Her voice softens: "No one can say with certainty what will happen in the future, but the apus and pachamamas [earth goddesses] say changes will happen in the earth. Not only in the Andes, but the spirits of the snows, the earth, and the mountains are united, and dramatic changes will occur."

I'm not sure what to say. I thank her and hang up the phone.

Outside my window, a red-shouldered hawk is sitting on a pine branch about twenty feet above the ground. I hold my breath. He plummets into the ferns and soars away, a black snake dangling from his beak.

3

FORTY-TWO DAYS

Naked

At six-thirty the next morning, I slipped out of bed and peered out the tiny crescent window in our bedroom at the park below. The autumn sun cast a few faint rays over the concrete paths, which were being scrubbed and polished by gardeners in maroon jumpsuits. I felt tired but hopeful about what lay ahead: My father had settled in peacefully after Señora Lucinda tucked him in last night. On the writing front, two days ago a British newspaper had given me the green light to do a story on Qoyllur Rit'i. Plus, I had been teaching morning English classes at UPC for three weeks now, and so far, the students seemed to like me. They were progressing steadily with their studies, and I was doing my best to fit into the university's highly sociable culture, keeping my introverted writer self under wraps. If I kept this up, my supervisor might allow me to switch to an afternoon teaching schedule next semester, and I could go back to writing in the mornings.

After getting dressed, I descended the angular, concrete staircase—built by hand, like all the old houses in the neighborhood, each step a slightly different length—being careful not to overshoot the final, narrow stair as I ducked under the low archway into the unlit dining room. My left hand automatically reached for the light switch, illuminating in one heart-stopping flash The Naked Man.

He stood frozen by the sideboard, stooped over in silhouette, his pale skin hanging in folds from his buttocks. He turned to me, and I saw his penis and a ponderous testicle sac dangling between his thighs.

A shock ran through me, as brutal and irreversible as a head-on collision.

Wordless, he shuffled forward, his eyes blank with terror as he glanced around the room.

I flicked off the lights and rushed to his side. His arm felt surprisingly frail as I guided him, step by halting step, down the tiled hallway to the room Señora Lucinda and I had made for him.

"It's all right," I kept saying in a quaky voice, the panic rising in me like bilgewater. "Everything's okay. You're in Lima now with me and Jorge. Your nurse comes at eight."

But as I helped him into his flannel bathrobe and convinced him to lie down on the bed, I kept seeing the thing a daughter isn't supposed to see, seeing them spiral in a slow, pendulous way that reminded me, oddly, of the Bavarian motion clock that used to stand on the mantel of our house in New Jersey, its gold balls twirling right, left, right.

I got through the morning teaching the students how to make polite requests in English. I was a convincing replica of a tough but caring *americana* professor—the persona that had gotten me hired in the first place—and all the while, I had a pain in my gut like I had been punched by a heavyweight fighter, a psychic wallop that reverberated so deeply, everything around me seemed distant and unreal.

My father was running around the house naked. My eccentric but highly dignified father, who wore suits to the doctor and chit-chatted to neighbors about Sir Francis Drake, had gone off his rocker and was prancing around in the buff like a loony. And now he was *our* loony to contend with, our burden. What had we gotten ourselves into?

A little after six at night—after spending the afternoon at a café in the airport, interviewing a climate change specialist for my Qoyllur Rit'i story—I opened the front door, dreading what I might find. The living room and the dining room were quiet and still; the mirror over the fireplace reflected a bouquet of fresh daisies in a blue-and-white vase, and there was the reassuring smell of chicken stew lingering in the air.

Relieved, I dropped my briefcase and followed the guffaws of a sitcom laugh track to my father's room. There, propped up with pillows in a La-Z-Boy recliner, dressed in a sweatshirt and sweatpants but barefoot, sat my father, watching an episode of *Friends*, with Lola curled on the floor beside him. He smiled absently as I patted his hand, and he returned his attention to Monica and Rachel arguing over a lost earring. Behind him, on the foldaway sleeper-couch, sat Jorge, looking grumpy and tired, and next to him, Señora Lucinda, in her white getup. She leaped up.

"Hola, Señora Barbara," she said, her soft brown eyes clouded with worry.

Everything in the room seemed calm and organized, yet I could sense an undercurrent of tension—in the rigid set of Señora Lucinda's shoulders as she counted out my father's pills, in the tentative way she held out the cup of water, calling, "Mister John, jor *pastillas* [your pills]." Her smooth, plump hands with their tapered fingers—previously so expert at weeding the garden and dicing onions for *ají de gallina*—now seemed hesitant and inept.

Deaf to Señora Lucinda's cajoling, my father stared at the TV, smiling as one of the characters tried to flirt with a pizza delivery girl. "There are other gas smells," Ross said lamely. "Methane smells . . . ." My father laughed several beats behind the laugh track.

"Was he dressed the whole day?" I said in Spanish to Jorge.

"Yes, we got him into clothes around noon. But"—he glanced at Señora Lucinda, a look of annoyance flashing over his face—"this one doesn't know how to manage him," he finished in English.

Jorge took the plastic cup from her hands. He thrust the pills in front of my father's rapt face. "John, here are your vitamins," he said firmly.

My father chased them down in one gulp.

We opened the daily log Señora Lucinda had started keeping on "Mister Jhon," as she spelled his name. Since she had arrived after 8:00 a.m., there was nothing about Mister Jhon running around naked, just an account in Spanish of *el paciente's* breakfast ("Kornflex" with milk, coffee, and a banana) and him getting dressed without a fuss. At nine, they walked in the park for an hour and then lunched on chicken and mashed potatoes. In the afternoon, he took a nap, watched TV, and talked in English (to whom it was unclear) and had dinner. All in all, a noneventful—that is, successful—first day for an eighty-six-year-old World War II veteran abducted from Florida and plunked down in the capital of a developing country.

When the sun started going down, however, it was a different story:

5:30 p.m. He doesn't want to take his pills. Nothing stays down. Stubborn.
6:00 p.m. Patient didn't obey orders. Patient very irritated.

I looked at the slippers thrown in the corner, my father's stubborn profile, the bottles of pills arranged in precise columns along the dresser top. Battlelines were being drawn. Should I be worried?

"He is disoriented by the move," Jorge said decisively in Spanish. "Things will be back to normal soon."

"Has he said anything about CB?"

"Nothing. Maybe he forgot about him in rehab."

"No way. That dog was the center of his life."

"You're wrong," said Jorge. "He's already gotten over it, see?" He nodded at my father, who was absentmindedly patting Lola's head. "He probably thinks Lola is CB."

No, he doesn't, I thought to myself. A laid-back, seventy-pound Labrador is no substitute for a screaming miniature poodle that likes to watch Humphrey Bogart.

"You're Fired!"

On Day 2, my father ate a banana for breakfast and lunch, refused to go to the park, and started sleeping on top of the sheets. He also stopped taking his glaucoma and prostate medications.

On Day 3, he ate more fruit, took off his clothes six times in one day, and said *No No No* to everything Señora Lucinda asked him to do.

On Day 4, he spent most of the day lying flat on his bed, eyes squeezed shut, hands folded over his ribcage like an elderly Dracula.

On Day 5, he stopped eating and drinking.

That evening we called in a doctor.

In the United States, the only physicians I had ever heard of making house calls were the doctors on *Leave It to Beaver* and *Lassie,* so I felt caught in a time warp at 7:00 p.m. when I opened the front door to find a short man in a white coat, carrying a bulging brown doctor's bag. In stepped a beaming, round-faced man of about thirty-five, with dark, curly hair and a little beard, who held out a hand to Jorge and introduced himself as Dr. Enrique Rodríguez. He turned to me and planted a kiss on my cheek, a familiarity I had come to expect with doctors in Peru (although I had never been able to get used to kissing my gynecologist).

He apologized for being late, mentioning the terrible Lima traffic this time of night, and all the while I was pinching myself that we had convinced a geriatrician from Peru's top Alzheimer's clinic—someone we found in the phone book—to make a house call on two hours' notice, for two hundred Peruvian *soles,* roughly the equivalent of seventy U.S. dollars.

Dr. Rodríguez's teddy bear–like face grew serious: "¿Tu papá no está bien, cierto? [Your father isn't well, right?]"

He said it in the exceedingly gentle, solicitous tone Peruvian doctors use when talking to women about anything medical—the diagnosis of a wart, a pregnancy, pancreatic cancer.

"Sí, mi pobrecito papá está muy mal," I said. (Yes, my poor little father is very ill.)

Dr. Rodríguez paused. "Can we go see your father now?"

I was surprised—the scheduling office hadn't mentioned the doctor spoke English.

A Mahler symphony was softly playing as we entered my father's room. Señora Lucinda was sitting on the couch, writing in the notebook. My father lay stretched out on top of the bed like Nosferatu. His eyelids fluttered as I turned up the light, but he squeezed them shut again.

I crouched by his bed: "Hi, it's me, Barbara. Wake up."

No response.

With Señora Lucinda's help, I propped him up in bed. He glared out of one crusty eye. "What's the big idea?"

"I brought someone here. A doctor. He's going to help you feel better."

"Hmmmph," he snorted.

The doctor came over. "Hello, Mr. Drake," he said in English in a jovial, high-pitched voice. "I am Dr. Rodríguez. Nice to meet you, sir."

My father spat on the doctor's outstretched hand and turned his back to us. Even I was shocked. "I don't need a doctor," he mumbled.

Dr. Rodríguez was wiping his hand with a white cloth handkerchief, visibly shaken. "Oh, no, Mr. Drake, that isn't nice," he said, his voice rising higher. "I am here to help you get better."

"I have to give him a little medical exam," Dr. Rodríquez whispered to Jorge in Spanish.

After some wrestling, Jorge and I got my father upright again. I was reminded of my father's childhood nickname: Battler Drake. He would box the other kids during recess and claim their lunch as a prize if he won. There wasn't always enough food at his parents' home in Easton.

Dr. Rodríguez stepped forward cautiously, brandishing a stethoscope. "Let me just listen a bit to your heart, Mr. Drake."

I caught my father's arm as he thrashed out at the doctor. For a frail octogenarian, he was surprisingly strong when riled up. My father glared at me and jerked his arm harder. Jorge stepped forward and got hold of his shoulder and wrist.

The doctor slid the metal chest piece inside my father's bathrobe and listened, frowning: "How are you feeling, Mr. Drake?"

"How am I feeling? How am I feeling? Oh, boy!" My father rolled his eyes. "I tell you, none of it works."

"What does not work?"

"This! The whole business you've got going here . . . . Get him away," he said to me.

"Just a few minutes more," I pleaded as Dr. Rodríquez felt for his pulse.

My father scowled at the doctor. "Are you crazy? You're not going to find anything there. I tell you, I'm dead."

Dr. Rodríguez laughed nervously. "Oh, Mr. Drake, you are not dead."

"I'm dead. It's all over!" His voiced keened at the high note of desperation I remembered from my teenage years, when he started drinking more.

"Please calm down," I said.

My father tore his arm away from Jorge's grip. "Enough of this. All of you, get out! And you!" He pointed a shaky finger at the doctor. "Who the hell are you? You're fired!"

In the dining room, Dr. Rodríguez reached into his coat pocket and drew out the handkerchief, mopping the beads of sweat on his brow.

"Whoo, whoo! 'You're fired!' He has quite a personality, your *papi*," he said, dabbing at his dark curls. "I cannot believe this. I have never had a client who didn't like me."

Really? I thought. I was sure that as a geriatric doctor, he saw lots of crabby oldsters, but I let it slide.

"Yes, this is the first time," he panted. "Not even at the nursing home I run in lovely San Isidro. Whoa! Okay, yes, let me see what we can do."

Dr. Rodríguez wrote out new prescriptions for the medications my father had been taking in the United States. He added one for the dementia drug Aricept and another for an oral medication called Risperidone, two drops, once a day. I asked what the Risperidone was for.

"To calm your father down," he said in Spanish.

"He's not eating or drinking. Can you help with that?"

"Don't worry. He needs time to adjust to his new environment. He'll be better soon."

"Really?"

"Of course. All he needs is lots of rest and love. You can do that, right?" He looked at me tenderly and patted my arm.

While we were settling the bill, he answered his cell phone. "Okay, Little Princess . . . in five minutes . . . yes . . . kisses," he said in Spanish.

He paused at the front door: "I have had many clients. And all of them liked me. Your father is the only one who ever fired me. The only one!"

He shook his head and disappeared into the foggy Lima night.

The Bossy One and the Fatso

When a family member leaps into an alternate dimension and refuses suste-
nance because they are convinced they're already dead, it helps to have someone
in the house who has experience with disasters—that is, the person I married.

Since my father arrived at our house and began unraveling, Jorge had
snapped into Red Cross mode. Any disaster, he claimed, could be managed if
you followed one simple principle: "delegate, delegate, delegate." Here in Lima,
that meant assembling a good team.

Problem 1: Father won't eat or drink. Call doctor. Check.
Problem 2: Father roams house naked and confused at night and tries to pee
   in windowsills. Get temporary night nurse. Check.

We got Maggy's number from a nursing home near the U.S. Embassy and
brought her on board the night after Dr. Rodríguez's visit. She was about five
foot two and built like a spark plug, with cropped black hair, rimless glasses,
and a low-key, androgynous manner. Her papers said she was twenty-six, but
to me she seemed ageless. I could imagine her looking the same at fifty, or
fifteen for that matter. In her relatively brief life, she seemed to have seen all
the trials of geriatric life—every stray turd, tantrum, and curse word—none of
which fazed her. Which was good because over the next few days, my father's
condition grew worse, not better.

He pushed away bowls of Señora Lucinda's heavenly soups with a scowl of
disgust; he spat out Gatorade on the floor and hid pills under the seat cushions.
Sometimes he snuck out of his room to make forays into the rest of the house.
I would come home to find round bite marks in the unpeeled mangoes on the
sideboard. Señora Lucinda claimed not to know how the bite marks got there;
I had an idea. She was probably eating my father's untouched lunch, along
with her own, and lying down for a siesta afterward. That gave my father the
opportunity to raid the fruit bowl.

Jorge was angry Señora Lucinda might be sleeping on the job, but I secretly
empathized with her. It must have been exhausting to care for a patient as
angry and unhinged as my father. According to her notes, he was now taking

off his clothes seven or eight times a day and refusing to take any medications, including the one that was supposed to calm him down. "The patient doesn't obey," she wrote in her round, schoolgirlish hand. The truth was, the patient had gone over to the Dark Side and was alternating between catatonic and violent states, like the little girl in *The Exorcist.*

I was scared and at my wits' end. What must Señora Lucinda be feeling? I could see now she didn't have the personality or skills to keep up with my father's continual fluctuations, let alone control them. She tiptoed after him with his pills and a cup of Gatorade, calling, "Mister Jhon, Mister Jhon," while he ignored her or shouted things in English that, if she understood them, would have broken her gentle heart.

"You goddamn bitch," he yelled, "leave me alone!"

A day later, that morphed into: "You're too fat for me. Get the hell out of my house."

Señora Lucinda laughed uncomfortably and twittered back at him in her low tones: "Oh, no, Mister Jhon. *Tranquilo, tranquilo.*"

My father kicked off a slipper; Señora Lucinda put it back on his foot. He kicked it off again; she put it on, and he kicked her in the shin. She held his hand to guide it into a shirtsleeve. He tried to slap her in the face. She told Jorge the hitting was normal, that some Alzheimer's patients did it, but I didn't feel right about it. Something awful was looming.

My father also called Maggy a bitch, but that didn't matter to her. She just ignored him and silently got the job done. If he wouldn't put on his pajamas, she held his arms down with one hand and stuffed his bony calves in the pants legs. If he spit out the pills, she crushed them with yogurt, added a gob of honey, and tipped the mixture down his throat. Maggy must have been half Señora Lucinda's weight, but faced with an *americano* a head taller than she was, she knew how to gain leverage. She also knew to wait and pounce. At night, she slept upright in an armchair in a corner of his room, under a blanket, ready at a moment's notice to guide him to the bathroom or give him the medications that fell at clockwork-like intervals. Nothing got past her.

Not that my father was thriving under Maggy's care or felt grateful. He actively disliked her—and suspected something was up.

"Who is that man?" he asked one morning as Maggy was leaving.

Who was that man? It wasn't that strange a question. As mentally confused as he was, my father had grasped the dynamics of his new living arrangement. He was stuck in a room with two irritating strangers: a silent, butchy type who bossed him around at night and a meek fatso who chased him around during the day. Everybody wanted him to take nasty-tasting pills and eat cooked food and keep his pants on 24/7, and from his vantage point at the tippy top of the cosmic Ferris wheel, all of us were crazy bitches.

Punched

Day 9, Day 10. A man was lying in bed. Eyes clenched shut.

"Do you want to get up?"
"No."
"Do you want to eat?"
"No."
"Do you want to watch TV?"
"No . . . leave me alone."

Secret forays into the kitchen and dining room to eat fruit. Half-eaten mangoes and grape pits left in the decorative ceramic bowl. Mealtimes: No, no, no. Shooting dirty looks at Señora Lucinda and her little cup of water. Did he guess it contained a drug?

Jorge brooded over how to get him to take the Risperidone. The answer came while watching a program on the Renaissance. He took a handful of green grapes and injected five cubic centimeters of tranquilizer into each. Six grapes = one twelve-hour dose. He arranged them in a bowl with some apples and casually left them on the sideboard. My father snuck the grapes throughout the day. Outfoxed by Lucrezia Borgia.

Before anyone else in the house ate a grape now, we had to check it for puncture marks.

What should we do about my father's refusal to eat? we asked Dr. Rodríguez. "Don't worry yourselves," he said in a soothing voice. "Your *papi* will be back to his peaceful, caring self soon. This is an adjustment period after the move. Relax."

My shoulders were tied up in knots. When I looked down to grade papers, a stabbing pain shot up my neck into my cranium.

On the afternoon of Day 10, I sat down to transcribe my interview with the climate expert and felt It for the first time since my twenties. The terrifying presence, like a massive shelf of brittle slate, teetering over my head. Dark, ominous. One sudden move, and I was done for.

That evening, I peeked into his darkened room. He was lying on top of the sheets, in a burgundy dressing gown and white crew socks. His mouth was slack, snoring.

I remembered that same mouth, twisted in reproach, in the late 1980s, as he sat perched on the sofa in Toms River, watching me plead with him. I had driven down from the city to finally confront him about what had happened at Purchase College all those years ago—him telling me I'd never make it as a writer, selling my oboe, leaving me nearly penniless in Manhattan, of all places. Why had he been so cruel, I wanted to know?

He gazed at me for a long time, a wounded look on his face. I took a sip of the strong coffee I'd brewed to fortify myself for the ordeal.

"Barbie, I never did those things," he finally said, his eyes filled with tears. "Never. I've always been a good father to you. I've always supported you in everything you do."

"No, you haven't! What are you saying?"

A tear trickled down his veined cheek. "I've always been a good father to you. Remember when I used to go to your concerts at Purchase? Wasn't that fun?"

"Arggh!" I screamed, hurling the coffee cup at the wall. It exploded by his record collection, neatly alphabetized, brown liquid splattering over Albeniz, Albinoni, Arnold . . . .

From that day on, my father had never budged from his denial, let alone apologized for what he had done. How could you make peace with a person like that, I now wondered? A person who hurt you repeatedly and later insisted he had done nothing, making you doubt your own sanity? It was a wound that never healed. I had cauterized it when I joined forces with Jorge, but now I had gone and done a foolish thing—through financial desperation, perhaps due to repressed childish hopes—and let my father into my life again. How stupid could I be?

These thoughts flitted about—bats banging into walls—as I looked down on my father in this strange room in this strange city in this strange land. The rift between us had never mended. I had done my best to pretend his opinion didn't matter, I had pushed on with my life and my writing, but secretly his influence lingered, like a curse uttered by a goblin in a dark fairy tale. It sounded completely overblown, of course, unless you were living it, like I was.

And now he was here, splat in the center of my life abroad, and my old neuroses were banging on the door. The teetering slate. Maybe the Other Thing would be next. God, no. I had worked too hard to go back to that.

I won't let you ruin my life, I thought, anger rippling through me. *I will not let you drag me backward.*

I went into the dining room and dug around in the drawers of the Italian server until I found the two skeleton keys. They locked the door between the dining room and the hall to my father's room. Until now, we had been propping it open with a rubber stop.

I caught Señora Lucinda as she was leaving the kitchen, bag in hand, to catch the bus home. Maggy, who had just clocked in, was right behind her, carrying a mug of hot tea and a plate of *triple* sandwiches, made with layers of egg, tomato, and avocado.

"One minute," I said, holding up the keys. I showed the women how to work the lock. "We will lock the door all the time now," I said. "So he doesn't escape."

"Yes, señora," said Maggy, popping a sandwich triangle in her mouth.

Sadness weighed in Señora Lucinda's eyes. "Yes."

"Call upstairs if you need help," I said. "Always keep the key with you. Give it to the other person when you change shifts."

Maggy disappeared into the hallway and shut the door behind her. I listened as the key poked in the lock a few times and clicked shut. I was surprised I didn't feel more relieved.

"Good night, señora," called Señora Lucinda as she quietly let herself out. I caught sight of her through the front window, her face turned upward to the streetlamp, which was flocked with tiny colorless moths. She stood there for a few moments, apparently thinking about something.

Upstairs, I began working on my transcript. I could feel the heavy slate looming over me. *Bad. You've done a bad . . .*

"No, I am not going to be punished," I said out loud. A glass bottle crashed outside in the park. Teenagers, probably. "I'm safe." My haggard face peered back at me from the computer screen. I was worn-out from the worrying—about my father, about money. It was costing a lot having a freelancer like Maggy here every night. Hopefully, my father would soon get over this rough patch and we could go back to having just Señora Lucinda during the day.

The rehab psychologist did say some people got over Alzheimer's, right?

The notes in Señora Lucinda's daily journal grew shorter and terser. "The patient won't put on his clothes for anything." "Won't obey. Escapes the room naked." She left the room for fifteen minutes to make a fish stew; my father peed in his bed. One day when she tried to change the sheets, my father "almost hit" her, but she "caught him by the elbow," she wrote. So, either that happened or he really did hit her, and later she wrote that he didn't. Whatever. I hoped it didn't happen.

A day after the "almost hit" incident, I came home around 2:00 p.m. and helped Señora Lucinda change my father's wet sheets. After that, a diaper seemed in order, so I hauled out the adult Huggies Jorge had brought on the airplane. I held down my father's arms while she slid the paper diaper under his butt and fastened it around his waist. In less than two weeks, I had learned not to be shocked at the sight of his naked body. It was amazing what human beings could get used to when they had no other choice.

"There, Dad," I said as he glowered at me. "Now you'll be comfortable."

"Just stop it. Let me die."

I let go of his arm.

His fist caught me square in the left eye. Everything went black, followed by white sparks, an intense blanketing pain.

My father just hit me, I thought as I covered my eye and fell to the floor.

"Señor Jorge! Señor Jorge," yelled Señora Lucinda. "Your wife is injured."

My eye socket and something behind my eye were throbbing. Was I going to lose my sight?

The floorboards creaked as Jorge knelt by me. "Are you okay?"

I tentatively uncovered my eye. Jorge's face and the legs of the bed were blurry. "I don't know."

Jorge helped me upstairs and into bed and gave me an ice pack.

"Do I need to call a doctor?" he asked.

"No. Just . . . leave me alone."

I lay in the dark room, listening to the taxis honking outside. A bowl of Señora Lucinda's squash and potato soup lay on the nightstand, untouched. He hit me. My own father had hit me.

I lifted the ice pack and cautiously touched my eyeball. It was chilled and convex—a good sign. At least it wasn't squashed jelly. I might not be able to see again through that eye, though. Better just to lie here in the dark and regroup.

He had never hit me. He had never hit my mother or any other woman, as far as I knew. That was part of his midcentury ethos. Battler Drake, defender of women and Uncle Sam and slugger of bullies and the Axis powers. Now I was his punching bag.

The tears welled up from some forgotten cistern. I could hear the waves crashing, pounding on the gritty sand, the seagulls calling. The roar got louder, tugging me back to a long-ago trip to the Jersey Shore with my parents.

I must be four or five years old. I am wearing my red one-piece with the stripes in front. The one with the baggy straps that slip off my shoulders.

Goosebumps on my legs, I'm wading into the cold water at Asbury Park, my father behind me in his blue swim trucks and upside-down sailor's cap.

Each time a foamy wave rolls in, he lifts me high in the air, like a ballerina, his sunburned arms around my tubby waist. "Up we go, Barbie." Sometimes the rough waters knock him back, but he stands tall and holds me tight so I don't fall. I can smell the coconutty lotion on his hard chest; I can feel his strong, muscly arms; I can hear the other kids screaming; I can see the seagulls dipping and circling overhead.

Up to the blue sky and white-hot sun, down to churning water. Up, down. Up, down. I'm flying, too, like a seagull.

"Daddy, again. Again."

He will never drop me. Never. I am certain of it.

The door to our bedroom cracked opened. It was Dr. Rodríguez, holding his doctor's bag in front of his tummy like a shield. "Mrs. Vera? May I come in and examine you?"

I flinched as he turned on the light. He sat next to me, bedsprings groaning.

"Look left, right, up, down," he said, shining a beam into my eye. "Does it hurt?"

"No."

"How many fingers do you see? Is it blurry?"

"Four, no." I started crying.

"Mrs. Vera," he said, "These things happen with Alzheimer's patients. Do not take it personally. It is not your *papi* who hit you. It is the disease."

"But isn't he going to get over it?" I sniffed.

He looked at me, puzzled.

"A psychologist in Florida told Jorge that when some old people get urinary tract infections, they temporarily get Alzheimer's, but when the infection is cured, they go back to normal. My father had a UTI in January, we think, but he's better now, so—"

"Ah, I understand what is confusing you," he said. "No, from what I have seen of your father, you should not expect his disease to go away. We can treat his symptoms, but . . . ."

A suffocating feeling came over me.

"Look, Mrs. Vera. He is very frightened right now. He is having the hallucinations; he does not know where he is. He needs . . . love."

"Love," I repeated. "But he's not eating. Shouldn't we do something about that?"

"First, your father needs love. And I am leaving these for you," he said, laying some brochures on the bedcover. "You need to learn about the Alzheimer's. But later. After your eye improves."

When he was gone, I picked up the top pamphlet and squinted out of one eye: *The 10 Warning Signs of Alzheimer's.* I draped the ice pack over the left side of my head and settled in.

"Are you better?" Jorge asked that night, stroking my forehead.

"A little, thanks."

"Don't look in the mirror yet. It's turning purple. I'm sorry he hit you."

"Me too."

He opened a paper bag lying on the bed and lifted out a glass vial.

"What's that?"

"CBD oil. From hemp. I got it from Henry's friend."

"Will it make me high?"

"No. It's not hallucinogenic. It will just make you very relaxed."

"Oh, whatever," I said, overwhelmed with everything. Punch, black eye, CBD. "Just put it over there." I motioned to the nightstand.

After he left the room, I curled up on my side, emotionally spent. It reminded me of when I was a kid. The feeling had started off as nothing and then gradually accumulated into Something Not Right and then, by fifth grade, into Something Really Bad. But in the early years of elementary school, I barely noticed it. It was like the long red hairs at the sides of my face, just out of my field of vision. Once I caught clear sight of the Bad Thing, though, I couldn't ignore it. Then the bad feeling was with me all the time. It was connected with him.

Now it was back.

I asked a fellow teacher at UPC to sub for me and emailed my supervisor that I wasn't coming in the next day. I told myself it was the black eye, but it was more than that.

The next morning, I stayed in bed dripping CBD oil on my tongue. And I let it come flooding back.

The Big Blue Table Redux

Age eight. I'm sitting at the big blue table in the dining room, chunky pencil in hand, writing a poem about a baby bird. On the wall to my left, above the sideboard with my father's stinky pipe collection, is a big map of Paris with all the little buildings and a big river in the middle. To my right, three tall windows with lacy curtains let in the fading afternoon light. I stare down at the wide-lined paper, an excited feeling stirring in my chest, word sounds singing in my head.

see / be, around / ground

sing / ring? bring?

I don't need to look back at the living room to feel my mother's presence; she is always there in her green armchair, reading her books, drinking her coffee, a pack of cigarettes on the wooden side table. Our beagle lies next to her chair, on the teal-blue carpet, snoring and farting. My mom is a secretary at Schering Drugs. She has cat glasses, like me, and short brown hair. My hair is long and red, in two thick braids. I want to read and know everything like her, one day.

"How do you spell *flutter?*" I call out.

"*F-L-U-T-T-E-R.*"

I chew on the end of my braid and try to feel like a baby bird sitting in its twiggy nest. It can't fly. It can't sing. Not yet. But one day . . .

I leave the finished poem on the table for my father when he comes home from work. He has a job in the post office. He's smart and strong and kind and can fix anything.

"Barbie wrote this?" he asks that night. He reads silently, his long thin fingers thrumming on the table.

> Baby bird, open your eyes and see
> What a wonderful thing life can be.
> As you flutter about and around
> See above you the sky, below you the ground.
> Though you are little and cannot sing.
> Deep down inside you're a wonderful thing.

"Oh, very nice," he finally says.

Pride bursts inside me.

My mother puts the poem on the refrigerator, next to my crayon drawings of farm animals and go-go girls in cages. A few days later, the poem is gone.

One weekend, I'm playing Barbies in the dining room. I peek in the bottom drawer of the sideboard. There it is, "Baby Bird." He has copied it out in his perfect handwriting with the puffy capital letters, along with some other poems I wrote, the originals stacked underneath.

He copied them out, so he must really like them.

I put my original of "Baby Bird" back on the refrigerator.

At the end of second grade, my mother counts the poems I've written this year. Seventy-eight. One of them starts, "I will be a writer and live by the sea."

"Keep it up, Barbara," my mother says.

She puts the poems in a Hess's department store box labeled "Barbara's Writing." The box goes on a shelf in the living room, under the Bavarian motion clock. The two gold balls spin back and forth, clockwise, counterclockwise . . .

Age nine. My third-grade teacher, Miss Wally, is a sadist. She plays favorites. I'm not one of them. Kids she doesn't like get yelled at or put in the corner. A boy with a clubfoot locks himself in the bathroom: "I hate her, I hate her," he screams. Rather than make a scene, I write a story in which mean Miss Wally goes on vacation to Bermuda and comes back happy and nice. I get sent to the principal's office and then to a psychiatrist in downtown Bloomfield.

I play with dolls and puzzles and answer questions.

"There's nothing wrong with her," I overhear the doctor saying to my mother afterward. "She came up with a creative way to solve her problem. We get lots of complaints about that teacher."

Returning to school that afternoon, I stare through the chain-link fence at the playground. My classmates are playing dodgeball; their shouts bounce off the asphalt, like the hard, pink rubber ball they are hurling at each other's heads. For the first time, I feel different from the other kids.

That night, my father tells me to "watch it" with Miss Wally.

"I know the type," he says to my mother, his face all red. "Puritanical bitches."

"John—," my mother says warningly. They go up the creaky stairs to their bedroom and close the door.

I am lying in bed, almost asleep, when the door cracks open. Light from the hallway silhouettes his crouched form, draped in a bathrobe.

"Grrrr," he growls, creeping forward. The long piece of hair he combs over his forehead dangles to one side.

"Tyger Tyger burning bright," he recites in his nasal voice.

"In the forests of the night," I chime in, as I have since I was very small.

"What immortal hand or eye / Dare frame thy fearful symmeTRY!" we finish together.

"Good night, Daddy." I hug him tight. "Hugger Bugger."

He squeezes back. "Hugger Bugger."

I keep writing stories and poems, some in class. At the bottom of the page, we sign our name in cursive and pledge we wrote it ourselves. Miss Wally corrects our spelling in shiny red pencil and sends some of our work to children's magazines.

Ages nine, ten. I'm in fourth grade now and in Mrs. Parkhill's class. Things are better. I'm getting fan mail from kids all around the country, and my mother lets me stay up late to write. My father visits me in the dining room and sometimes kisses me on the top of my head, sometimes not. He circles the big blue table, counterclockwise, as I pen odes to Martin Luther King Jr., the glory of English muffins, the Statue of Liberty. I feel the air stir as he passes my chair, leaving a faint charge, but I ignore it. The word sounds in my head are stronger.

My mother still sits in the living room, calling out spellings when I ask for them. Lately, she's been reading a book by a Russian guy with a beard who went to jail in Siberia. Sometimes she sighs loudly and goes outside to smoke.

As I write, Paris looks down on me, each tiny building sketched in precise detail. A bright green dot marks the apartment where she and my father lived for two years in the early 1950s, near the Luxembourg Gardens. He was a poet and a writer. She was a secretary for the Marshall Plan, whatever that was. Something to do with the government. He wrote a novel, but it didn't get published. Afterward, he and my mom moved to New York City, where he and a friend started a market research company, Packaged Facts. She worked at the United Nations and Columbia University and some other places. I was born ten years later, after he went to work at the post office.

Fourth grade keeps churning on, one weekly spelling test and several poems at a time. Then comes the night when he laughs at *squished*. I can't name what went wrong. But I feel it in my stomach.

Bad. It. Me. Something.

— ✦ —

I announce in my diary that from now on, I will get perfect grades in everything.

It's easy. I've been doing it since first grade. Only now I get nervous before tests. I develop headaches. Mrs. Parkhill tells me to relax. I ignore her.

My father praises my report cards, all As and Os for Outstanding. "Barbie, you're the Greatest!" he says, giving me a big bear hug: "Hugger bugger."

"Hugger bugger," I say, hugging him back extra hard, my heart full.

I. Am. Perfect.

— ✦ —

The Hess's box fills up with my writing. My mother starts another.

— ✦ —

Age eleven. I'm writing longer stories now and plays. My fifth-grade teacher praises me nonstop in front of the other students. A gang of girls corners me in the coatroom.

"Stop being so smart," the ringleader says. She punches me so hard I bang into the wall. The metal coat hooks dig in my shoulder blades. "Stop skipping down the halls."

I stomp on her white go-go boots and run to the bathroom. *Idiots,* I think.

In March, I get my first-ever word wrong on a spelling test. *Vacuum.*

"I'm a failure," I sob to my mother.

"Don't get so worked up, Barbara. You'll get a headache. Here," she says, handing me a tissue. She pulls an envelope out of the Samuel Beckett book she's reading. "Oh, and this came for you."

An acceptance letter for another poem and an award from the Junior Mc-Call's Club. That makes me happier.

My father says nothing about the Junior McCall's award. His eyes go to the spelling test on the coffee table. "Only a 90 out of 100?" he asks.

I hate myself, I hate myself, I think.

I find it in the laundry room in the basement, high up on a shelf over the washer and dryer, next to his fishing tackle. "Take the Dubious Road."

I climb down from the dryer and blow the dust off the old green-and-black cardboard box. The sides are falling apart. The corners have been taped over many times. The masking tape is brown and crumbling, like something out of King Tut's tomb.

Written on the inside of the box in my father's hand: "John Drake, 206 East 73rd Street, New York, NY. 2nd copy. A Novel."

The box is filled to the top with sheets of yellowing onion paper, neatly typed. The last page is numbered 476.

The top sheet says, "PLEASE REPLACE MS CAREFULLY (will call for)."

I sit on the cold linoleum floor and read the introduction.

"This is the account of how one young man, having prolonged his adolescence, attained manhood and gave direction to his life. It is a typical American story, for up to the time he attended college, including the little he saw of the war, he was a dreamer—*not really a man*. And he idealized everything. Money was the power of darkness. People lied, stole, cheated, were greedy and suspicious of each other because poverty continually menaced them. Religion was the healing balm of the woes of living."

Blah, blah, blah. I flip forward to find the good parts. In chapter 5, Leonard, the main character, meets Diane at a college dance in his hometown of Easton.

"She wasn't very tall, a bit over five feet, and she was small boned, with fine hands and graceful, childlike shoulders that drew looks of admiration when she wore her favorite, off-shoulder dancing gown."

Holy cannoli. It's my mother. Diane = Ann. Leonard was the name of my father's younger brother, who died of appendicitis when he was a teenager.

I step into the other room in the basement and honk a few notes on my clarinet. I'm supposed to be practicing.

Back to the manuscript.

". . . as they danced—he could smell her honey-blond hair—she had pressed her petite self so close to him to fit, as if by magic, as if by supernatural design, against the hollow of his shoulder, his lap; they had sipped the most exotic drinks, Orange Blossoms, Singapore Slings, Pink Ladies and Parfait Amours, and babbled like children; and when they kissed good night they had discovered— it was like a sudden electric shock—that they, their lips, were fated to thirst for each other forever."

Fated. Wow.

At supper, I stare at my mother's hair under the kitchen light. It is pale brown, ashy. But *honey blond* sounds nicer.

For the next few days, I alternate playing scales with reading "Take the Dubious Road" in secret. Leonard takes long walks in the country and writes some poems about nature. Leonard rails against his "narrow-minded, petit bourgeois family." Leonard finds his studies "meaningless" and quits college. Leonard and Diane plot their future life together—"travel, a free life, risking it all together."

I'm looking forward to some exciting scenes in Paris when, on page 51, Leonard embarks on—a job hunt in Easton! Seventy pages later, a "toughened old newspaper man" with a chewed-up cigar hires Leonard to be in charge of delivery routes for a newspaper.

The next three hundred pages are all about the business of newspaper circulation: how contracts are signed, how payments are made, how the delivery guys compete to get customers. So many details, so many dollar amounts. I stop following the plot and rifle through the rest of the pages. On 421, Leonard and Diane get married. Her hair is still honey blond.

On 440, on their honeymoon: "Through her wedding skirt he held her hips, he gripped the voluptuous, outside flesh. He undid the button at her waist and slid the fabric down. He had her by the hips, he had her by the thighs. He had her by the right of all that's young and pure and fresh—her skirt came tumbling down into a heap, she kicked her stockings onto that heap—he had her by the insides of her thighs."

I try to picture it. He is grabbing the inside of her thighs? *What* "had" her there?

"He was all man for her—she thrilled at that, she worshipped that—and she was all woman for him."

That gets my attention.

On the very last page, Leonard quits his job and announces he and Diane are moving to Paris. Au revoir.

I carefully straighten the pages, fit them in the box, and close the lid, bits of old masking tape flaking off. My father's novel is boring, apart from the romance and sex stuff. He should have made it all about Paris.

I feel sad. I wish the whole thing was good. I'm also mad. I push that feeling away.

Three words form in my head.

How dare he.

Age twelve. I write every day in the dining room, at the long grayish-blue table whose surface shows every precise brushstroke made by my father long ago. I play the clarinet, I take ballet classes, I look happy, but things are shifting.

My friends don't want to play with Liddle Kiddle dolls anymore. My friends want to wear eye shadow and lip gloss and hang out at Wright's Field and watch the sixth-grade boys play basketball. Yuck. What a waste of time. Makeup is stupid.

My best friend dumps me for another girl. Fine. I'll play Liddle Kiddles on my own.

My dad now has more drinks when he comes home. Three drinks. Four. I stop counting. He hasn't tucked me in for years. "Tyger Tyger" isn't a heartbeat thrumming through the forest anymore. It's just black words on a white page. That's okay. Writing still feels like the best thing in the world.

One night as I'm drawing cartoons in the dining room, I hear a thump behind me. My mother is lying face down on the teal-blue carpet, her glasses broken.

I put my hand on her back; I can feel her breathing.

"Mom?"

"Yes?" she answers groggily.

Our beagle is whimpering. I'm scared.

A few days later, my father tells me: "Your mother had a nervous breakdown. She's going to need our help from now on."

I accompany her every Saturday on the bus to Manhattan to see the psychiatrist. I sit in the waiting room, reading paperbacks of *Peanuts* cartoons she buys me at the Port Authority. When her visit is over, we go to a store on Lexington Avenue where they have black glitter spiderwebs and Day-Glo Peter Max puzzles and copies of *Ms.* and *Mad Magazine* and another one called *KIDS*, written by kids like me. My mom and I still laugh at the same things, but as we wind through the New York streets, my arm locked with hers, I worry she's going to get run over by some crazy cabdriver.

"Don't worry, Barbara; I'm going to be okay," she tells me through gritted teeth as she eyes the busy crosswalk. She and I are wearing navy peacoats and matching red mittens we made together in mother-daughter knitting class, our first pair. We didn't know when to stop knitting the thumbs, so they're as long as the finger part.

"Hop to it, kiddo!" she says, gripping my arm tighter when the light turns green.

Six months into going to New York, I pick up the latest *KIDS* and find the first chapter of a book written by a girl my age, named Ally Sheedy. "She Was Nice to Mice" is about a mouse at the court of Queen Elizabeth I. I'm so jealous, I feel it in my intestines.

That's what I'm going to do, I think, *I'm going to write a book.*

Age twelve. My mom keeps working at her secretarial job, but she's more tired now. She goes to bed after supper, around seven. I'm alone at the big blue table, writing, writing. I miss being able to ask her how to spell things.

My father, in his three-piece suit and silk tie, circles the table. Around and around like a shark. My body tenses. He stops behind me. I feel his eyes on my paper.

"What are you writing, Barbie?"

"School stuff," I lie. It's a story about Zeus and Hera and Mercury fighting for control of humankind.

I pretend-yawn and cover the dialogue with my forearm. I don't want him to inspect my handwriting, a messy blend of regular and cursive. Sometimes he pokes his finger at it. Like he does with my hair. He says I shouldn't let it fall over my face. It's not attractive for a young lady. I should use barrettes. What does he know? When I grow up, I'm going to live in New York and have lots of lovers and never get married. Like Gloria Steinem.

"You used to write lots of poems," he says casually.

"Not anymore."

"Oh really?" The words leap out of his mouth.

I don't like these conversations. No matter what I say, I can't win. I want to please him. I don't want to please him. I don't know what I want.

Bit by bit, I learn what happened to my mother. Her mother, Charlotte, had a breakdown when my mom was eight and was put in a sanatorium in Pennsylvania. My mom's father told her and her two sisters that Charlotte went off to become a nurse. My mom had to pretend Charlotte didn't exist. She and her sisters got a new mother, Ruth. But she missed her old one. One day my mother couldn't keep on pretending. She had a nervous breakdown.

Now my mother wants to find Charlotte again. She's still alive.

I feel upside down about it. Grandma Ruth isn't my real grandmother. My grandfather told lies. My mom had to pretend for so long everything was okay, it made her pass out on the living room carpet. My beautiful mother with the golden heart who smokes too much.

If my mom had a breakdown and her mom had a breakdown, does that mean it will happen to me too?

I dance, I play the clarinet in recitals, I get all As, my father *blah blah blahs* about how great I am. As long as I keep walking the tightrope in a perfect straight line, I won't fall off.

"Get your hair out of your face." My father's voice.

I start wearing a navy-blue lumberjack's cap, my hair parted in the middle and slicked behind my ears. I feel like Me in the cap, like I do in my faded Levi's and my oversized plaid shirts and my suede Wallabees, just like my mom's.

"Stop wearing that. You look like a tomboy."

Headaches, on the right side. In the back of my skull.

I retreat to my bedroom and write my stories there, lying on top of my flower-print bedspread in my lucky writer's shirt and the lumberjack cap. I don't leave my story ideas out in the open; I scribble them on scraps of paper, fold them into little squares, and stuff them inside a pair of baby shoes that hang from my bedpost. No one will think to look there.

I write and illustrate a detective story, "The Mystery of Art Smart & the Missing Birthday Cake," and mail it to the address they sing on TV: "Z, double O, M, Box 3-5-0, Boston, Mass. 0-2-1-3-4." They send me a release form, and I sign it.

Cap pulled down to my eyebrows, I dream of writing something completely new, something that will bend time—science fiction maybe. I tentatively plot out novels, rip them up, start over. I'm going to be more famous than Ally Sheedy. I have to hurry, though. Time is running out.

Sixth grade turns to the summer before junior high.

A Saturday morning, July 1973. Dappled morning light plays on my bedspread, yellow and orange flowers, celery-green leaves, mingling with shadows from the tall maple tree outside my bedroom window, and I feel so alive, so alive— bursting with delight in what my brain can do.

No more headaches. I've done it. I've come up with the idea for my first science fiction novel.

Twelve-year-old Paula from Bloomsbury, New Jersey, pierces the fourth dimension and travels back to 1685 to rescue the dodo bird from extinction. But when she returns to the present, she is hunted by a secret government agency that wants her—and her two living specimens—wiped from the Official Record.

I leap off my bed and run down the staircase, taking two stairs at a time.

My father is crouched at the landing in his upside-down sailor's cap, patching a strip of wallpaper where a guest at a birthday party left a greasy icing handprint.

I'm feeling so great, I don't stop myself. "Guess what?"

He puts down his roller and listens as I tell him the plot. His lips flatten. He blinks a couple of times like something's in his eye.

"That doesn't sound like much of a story," he says when I finish. "Who would want to read something as hackneyed as that?"

He goes back to wallpapering.

The full-of-light feeling goes out of me. In its place roars something I have been trying not to feel for years.

I run upstairs, slam my bedroom door, and rip off my hat. I rub my hair until it stands out like a giant red afro-halo.

"Argggh!" I scream at the wild person in the mirror. I flop on the bed.

Later that day, the roar is gone. The headache is back. I pick up the Snoopy pen. It feels clunky in my hand. Whatever. I have to get it down on paper.

"For my biographer," a hopeful little voice in me says.

"Shut up. You won't have a biographer," snaps another voice that surprises me with its viciousness.

"But it's my swan song," the little voice pipes up.

"Go ahead and die like Pavlova, then. Get your swan costume ready."

"Dear Someone," I begin, my hand shaking, "My father finally said it. 'That doesn't sound like much of a story.'" And with that octet of words went all my hopes, my feeble dreams, the small grain of chance that my writing would be more than mere scribbles on lined paper. My kind, admirable father, who with those words has extinguished the fire, the torch he handed to me.

"Perhaps it is envy, envy that one day I would not get a rejection slip like he did. I do not know. Although he said it regarding my future book, I knew it stood for all I previously wrote."

I look at what I've written. I know it's florid and overdramatic. I don't care. I just lost Me.

The next day, I wake up and pull my hair off my face into a ponytail.

I gather all my writing—my bedroom manuscripts, the Hess boxes, my *McCall's* and *Red Cross* clippings, the letter from *Zoom*—and dump it in a used postal bin. I shove the bin in the back of my closet, with the Liddle Kiddles and the knitted blue cap. My heart feels like a knot. A stupid old piece of gristle you'd chew and spit out on a plate because it's worthless.

I slam the closet door.

Writing is for children, I tell myself. I will never, ever, write again.

Age thirteen. Listening to classical music on the radio one afternoon, I hear a piece for soloist and orchestra that transfixes me.

It starts with a low throbbing in the cellos, followed by a soft melody in the violins that creeps steadily forward, like mist along a forest floor, secretive, more felt than heard. The mist flows over moss and tree roots, under heavy pine boughs, to a deep, deep clearing, which is pierced by a sudden light—the sound of an oboe.

The oboe sings very simply, buoyed by the beating of the strings. It says we love and glimpse beautiful things when we are children, and enchantment comes as naturally as breathing. And as our growing selves rush to meet this enchantment, it vanishes, like mist, and all that's left are crumbling leaves; dry, brittle twigs underfoot; clear, gray morning light. Quiet death.

The oboe sings this lament so achingly, so purely. The sound penetrates some hidden river inside me. It is saying what I can no longer say in words. It is beautiful but, above all, profoundly sad.

I sit transfixed through the remaining movements, waiting to hear the composition's name: *L'horloge de flore* (The Flower Clock), by Jean Françaix. The oboist is John de Lancie.

I am going to learn to play that instrument, I vow. I am going to give whatever is left of myself to that sound.

My illustrated story airs on *Zoom* a year later, after we move away from Bloomfield. Watching it is surreal. There are the magic marker drawings I labored over; there is my detective story being read out loud by the Zoomers. The husky-voiced boy I have a crush on plays the role of Art Smart, Private Eye. I love seeing my story come to life on television but feel embarrassed.

I wrote this as a child, I think. Embarrassment curdles to shame.

I tell no one at my new school that I have a story on *Zoom*. I don't tell my father. Only my mother watches the episode with me. She records it on the same Panasonic tape recorder we used to record Nixon's resignation speech the summer before.

The Land of No-Desire

Two days after the punch, I woke up around 8:00 a.m. It was a Friday, so thank-fully, I didn't have to teach. When I went to wash my face, I was startled by the purple-eyed monster in the mirror. I could almost make out the bony imprint of his knuckles below my eye socket. Rage boiled up. *How dare he.*

Downstairs, Señora Lucinda was preparing coffee and oatmeal, which she pronounced *quacker.*

"Ay, Dios mío," she said, putting a hand over her mouth.

"I'm okay," I said to her in Spanish and went to his room, determined to be tough.

He was curled on his side facing the wall, in a blue sweatshirt, diapers, and nothing else. His pills lay scattered on the floor, near a puddle of Gatorade.

"I have to talk to you," I said loudly. He just lay there. "Did you hear me? You fucking punched me."

"Oh," he moaned. "Oh, oh . . ."

"It's not okay that you hit me. I'm your fucking daughter, you, you—"

"I'm dead," he howled. "We're all dead."

The sound of his voice cut right through me.

I stood there breathing in and out.

My god. That shell of a person on the bed was what was left of my father. He was no longer Battler Drake. He was just a human being wracked by suffering.

I sat down on the couch, drained of all anger. In its place rose an odd cer-tainty: This dark place he was in—he had been here before. He'd written about it—a poem from his twenties, sometime after he got discharged from the navy. He had gotten typhus at Pearl Harbor and almost died. "Elegy III," it was titled. It was one of only three poems of his that I had ever read. He used to keep the poems in his middle desk drawer, in Bloomfield, along with black-and-white photos of himself and other sailors in Hawaii. As a girl, I would sneak into his office and look at the pictures: handsome young men, shirtless, flexing their muscles in the blinding-white sun. In some, my father's eyes were closed.

I could still recall it from memory.

I was sick
and went into the land of no-desire
caring not for food or drink
or the boundless joys that earth can give,
caring less for pursuit or pleasure
and still less for love.

Long days passed
in an endless string before my elephant eye.

I was sick
until I left the land of no-desire,
and lo! I began to live once more.

What in the world are we going to do about you and your elephant eye? I wondered, surveying his bony frame. He seemed a bundle of disconnected parts. A calf, a heel, a waxen hand. And somewhere in there—a brain growing less and less tethered to reality. The consciousness was there but shrinking, regressing. Like Qolqepunku Glacier. Melting into thin air. Creeping up the dark mountain. Leaving a messy trail of shards and mud and matted ichu grass from god-knows-when. And me, staring upward at the vanishing ice. Helpless to do anything about it.

My father had Alzheimer's and was never going to get better. I understood that now, thanks to Dr. Rodríguez and the pamphlets. He was our *carga* (burden), as they said here. Well, there was only one thing to do.

I touched my bruised eye. The familiar-looking stranger on the bed had done this to me. He was scared and wanted to die. I wouldn't let him.

Donkey Belly's Gray

Day 14. Sunday morning, ten o'clock. I was in the living room, peering out the front window at Parque Leoncio Prado. The sky was its usual wintertime overcast hue, a pale gray known in Spanish as *panza de burro* (donkey's belly), yet I could sense a brightness trying to pierce the cloud cover. Perhaps by midday we would have sun.

Since eight, the park had been stirring with activity: maids in blue-and-white uniforms walking their employers' dogs, smock-clad nannies pushing babies in strollers, children on bikes chasing pigeons, teenagers kicking a soccer ball around. Soon the couples would arrive, young people from the poorer districts of Lima who, like many Peruvians in their twenties, lived at home with their parents and sought anonymity in Miraflores's well-tended public spaces. Around 2:00 p.m., the sidewalks would be filled with families walking off their big Sunday *almuerzo*, typically with a bundled-up grandparent in tow. Wrapped in layers of sweaters and knitted scarves, *abuelo* would often be pushed in a wheelchair, its wheels making a slow metallic squeak as the family chattered away above the elder's head.

Just a few weeks ago, I had been imagining Jorge, my father, and me taking part in this Sunday ritual. Not today. My father was lying in a diaper on top of his mattress, passively trying to kill himself, while we waited for the arrival of a freelance nurse with a hypodermic needle. The money was pouring out of our household like a sieve.

I let the curtain drop, a somber chill returning to the room.

Jorge came down the stairs: "The RN will be here soon. Dr. Rodríguez has prescribed a shot of Haldol. It's a sedative."

"Do you know if it takes effect immediately?"

"Rodríguez says it will take a few days. But it should stop him from hitting people." He looked at my black eye, now yellowish green and brown. "Sorry, Barbara."

As I stepped into my father's room, my eyes were drawn to the large water spots on the sheets, signs of an earlier struggle. My father lay on the edge of the bed, pale as a ghost, eyes closed. Maggy, who was filling in for Señora Lucinda, gave me an update. It was as I expected: He had eaten nothing since he woke up. Two days with no food. A person could not survive like this. She held up the glass of diluted Risperidone my father had refused earlier—did I want her to try to force it down his throat? No, I told her. Someone was coming to give him Haldol.

"Ah," she said, visibly relieved. "They use that at the nursing home."

"Does it work?"

She gave me an inscrutable look through her glasses. "It is very strong, señora."

Half an hour later, Maggy, Jorge, and I lined up at the foot of my father's bed as a tall nurse in green scrubs prepared to give him an injection.

Manuel's smooth brow creased as he held the glass bottle of Haldol upside down and drew the clear liquid into the needle-tipped syringe. Five milliliters looked like a lot of medicine to be giving an emaciated senior citizen.

Manuel glanced at my face. "I am fully qualified to do this," he said in Spanish. "I know all about his sickness. It happens to many people. There are many different kinds."

I wondered what Manuel had seen. He had spent the last five years in Madrid caring for people with dementia, returning to Lima two months ago. From what I had been reading lately, he would have no problem getting work in Peru; the country was seeing a surge in Alzheimer's cases, just like everywhere else on the planet. I watched Manuel flick the air bubbles out of the syringe; his hands were large and meaty. He could definitely take on my father.

Needle ready, Manuel approached the bed, an alcohol-soaked cotton ball in one hand.

My father's eyes popped open. "Don't give me that thing," he said, backing up against the headboard. He raised his fist.

"Help me," Manuel barked in Spanish at Maggy. "Face down [*boca abajo*— literally "mouth down"]."

Maggy rolled my father on his side to flip him over.

"You don't understand," said my father. "I'm dying. Just let me die." He flung out an arm and smacked Maggy in the face. She didn't even flinch. Her cheek blazed red.

In the end, it took all four of us to hold my father down. To cushion the blows, we put socks on his hands. Brand-new navy Goldtoes from JCPenney. The needle went in his right butt cheek.

"You motherfuckers," he screamed.

The drug, a potent antipsychotic, would reach the maximum concentration in his blood in six days' time.

The Wasteland

Day 15. I was on campus, rushing between classes, when a Cusco number lit up on my phone. Shoot. I had forgotten.

"Señora Barbara, we go to Qoyllur Rit'i?" he asked in Spanish.

"Uh, probably."

He got agitated. "But you said yes. You and Señor Jorge will go in June—"

"Yes, Paco, we will go to Qoyllur Rit'i. But we don't have a lot of money this time. How much do you want?"

He rattled off a number. It was the equivalent of what the repairman in Florida wanted to re-overhaul my father's six-year-old HVAC unit. The new tenants were complaining it wasn't working properly. We had already spent fifteen hundred dollars to fix it three weeks ago. So far, we were losing money renting his house.

I ducked into the bathroom to inspect myself in the mirror after we hung up. My right eye was twitching. You could see the bruises around the other eye through my makeup. It all was so fragile: Paco's crappy Spanish, my crappier Spanish, the two of us trying to plan a trip to a melting glacier that would be gone in fifteen, twenty years. All this for a $250 newspaper story people probably wouldn't read. Paid for with a patchwork of funds that were rapidly disappearing. While my father's brain disintegrated. The whole flimsy edifice could collapse any minute.

This morning, I had taught my intermediate students about the conditional verb tenses—*might, may, could, should*. Too bad I didn't know those tenses in Spanish. Too bad Paco didn't know them either. But for this trip, he wouldn't care. He wanted certainties.

If the expedition did fall through, I knew enough basic Spanish to convey that to him accurately: *Unfortunately, Paco, we are not able to go to Qoyllur Rit'i this year. A thousand pardons. We cannot go.* But I would not be able to stop picturing his moss-covered hut, the charred firepit inside, the guinea pigs running around, his scared-looking wife, his four skinny children, his mean-faced mother stomping through the doorway, shaming him for losing out on this year's soup packets and strawberry jam.

Somehow, for everybody's sakes, I had to make this happen.

— ✦ —

That afternoon, I visited my father in his room. He was lying under a bright-blue Florida Gators blanket, finally wearing pajamas. Since getting the shot of Haldol, he had been sleeping more and drinking apple juice, said Señora Lucinda. But he still refused to eat.

I tugged on the blinds to let in more light. A fat brown *cuculí* landed in the dusty fig tree outside; "ku-KU-ri, ku-KU-ri," the bird called. The cheerless melody matched the mood of the room, which only got sunlight in the summer months, December through early March. Now it was almost April, and the sage-green walls Señora Lucinda and I had painted a few weeks ago were turning a flat slate-gray in the cool light. Wrong color choice, I now realized. My father was chronically sensitive to lack of sunlight, something my mother had alerted me to as a girl. When they lived in The Hague, she said, the constant cloud cover had made my father depressed. He took long walks and grew a beard. He didn't crack a smile in any of their Holland photos, not like the earlier ones of Paris, where he looked drunk with happiness and the wide French boulevards and budding flowers in the Jardin de Plantes. My little mother next to him, blushing, in her full New Look skirts and patent leather belts cinched around her tiny waist.

I pulled a chair by his bed. His face was colorless, paler than *panza de burro*. His cheeks were alarmingly sunk in. He had rapidly gone downhill in the last twenty-four hours. I was careful to lean back, out of range of his hands, which lay open on top of the blanket.

"Hey, it's me, Barbara," I said quietly. "Wake up." After a few seconds, I lightly brushed his wrist with my fingertips. His skin was crisscrossed with thick blue veins; blood was under there, pumping.

His eyes, when they opened, were a pale blue-gray, rimmed with red. He looked immeasurably sad but lucid. I felt like I was back with my old father.

"Oh, hi, Barbie," he croaked.

"What's up?"

"What's up, doc?" he parroted back.

"I understand you're feeling a bit under the weather, so I wanted to check on you."

"Hmmmph," he laughed, in what seemed an ironic way. He tried to get up on one elbow.

"Here, let me help you." I propped him up with pillows and pulled the Gator blanket over his lap. He reached out and patted the cotton fringe.

He looked at me for a while. Since he had been hospitalized, the spaces between his thoughts had grown larger. I had no idea what was going on up there. Maybe he was getting ready for another punch.

His fingers began to knead the fringe of the blanket.

"So, anyway," I said, "we're going to have to get you feeling better. Señora Lucinda—she's your nurse—she says you're not eating anything."

"My . . . nurse?"

"Yes. She's here today."

"There was a man . . . ." His fingers kept kneading.

"That's Maggy. She's not a man; she's a woman. Her name is Magdalena."

"Magdalena," he repeated.

I noticed a bowl of red grapes on his nightstand. "Want a grape?"

He shook his head no.

"They're good." I checked one for needle marks and popped it in my mouth. The skin was dry and slightly tart, the inside sweet and firm, not gooey. I cast around for something to spit the pits in and saw a box of Kleenex on the floor. Reaching for it, I noticed small brown pellets scattered by the baseboard; my father had been eating the grapes and surreptitiously spitting out the pits there. It was the sort of smallish-ly defiant thing a boy of seven or eight might do. At least he was eating something.

I cleaned up the pits and put the Kleenex back on his nightstand. "When you eat grapes, spit the pits in a tissue, okay?" I told him.

"I didn't eat any grapes," he said, mouth trembling. His fingers kept up a steady rhythm, rolling the blanket fringe back and forth.

"Dad, I know you spit the pits on the floor—"

"I didn't eat any grapes. . . . There's no food in this house." His fingers stopped kneading. He sank back into the pillow and closed his eyes, as if keeping them open was an effort.

His next words were so soft, I had to bend close to his stale breath to hear.

"I think . . . I think I'm going crazy," he whispered. "My mind. I can't stand it."

The anguish in his voice cut right into me. He knew what was happening to him, to his brain. How many weeks, months, years, even? I had a sudden flash from last December, when Jorge and I had flown to Gainesville, of my father sitting at the kitchen table, hunched over a game of Solitaire. One day he uncharacteristically spent the morning and afternoon in his pajamas, frowning at the cards. Perhaps he hadn't been playing. Perhaps he had been trying to remember how to play.

I think I'm going crazy. It was the sanest thing he had said in a long time. What was I supposed to say back? "Yes, you are losing your mind"? "No, you are not losing your mind"? What did you say to someone with Alzheimer's?

He had said plenty in the past to devastate me. I could do the same now. The rehab people never told him he had the disease. I could do that now. Just three little words. It would be so easy. I could say, "You have Alzheimer's," and he would understand because, for some reason, the fog had temporarily lifted in his brain, I could sense it. He wouldn't have the strength or the wits to fight back. It would sink into his consciousness, stain it, like a Rorschach blot. *This is you. This is what you are.*

The idea glinted at me, black and jagged, a streak of coal deep in my heart, tempting me.

A *cuculí* landed on the windowsill, breaking the spell. No, I decided. I would not seek my revenge on him. No one deserved that cruelty. I had to do the right thing by him.

I turned to me father. "You're not going crazy. You just had a bad urinary tract infection. We're going to help you get better. You want some water?"

He shook his head and sighed, turning toward the wall.

"T. S. Eliot was right," he said in a faint voice. "'This is how the world ends. Not with a bang but a whimper.'"

Dr. Rodríguez was still chipper when we called him. "Give it another day and then call me," he said, sounding distracted. Merengue started up in the background. "Come on, *chicos*, let's dance!" shouted an enthusiastic-sounding young woman, probably an *animadora*. They were professionals hired to liven up—

animate—events like children's parties and product giveaways at supermarkets. They made me cringe.

The next day, my father refused to eat and drink. I tried using a straw to dribble water into his mouth. He let it run down his chin. "Let me die," he wheezed.

I ran into Señora Lucinda later in the kitchen, dolefully stirring a batch of squash and lima bean soup. Her eyes were wet. We were both afraid.

"We can't go on like this," I said to Jorge, tracing circles on his bare chest. "He hasn't eaten for three days."

We were upstairs in the bedroom, and as much as I was trying to keep the "downstairs" separate from the "upstairs," it was no good. My father's crisis had certainly put a damper on our sex life. We no sooner started kissing then we were back on the topic again.

"Maybe we should go against Rodríguez and force my father to eat," I continued. "But is that right? I mean, if he doesn't want to, then maybe we should just let him—you know. He keeps saying that's what he wants. Oh, god, this is nuts. What do you think?"

Jorge pushed my hand away and sat up. "Fuck it. I'm calling the other doctor from the Alzheimer's clinic."

"What good will that do?"

"Jesus Christ, we can't give up."

I started crying. "We brought him here, and he's going to die, and it's my fault."

"Shhhh, it's ringing."

"What's this one's name?" I asked.

"Cera. Like *wax*, in English."

We got him on the fourth ring.

Doctor No. 2

April 7. Exactly eighteen days since my father's arrival at Jorge Chávez Airport. It felt like eighteen years. Today was my fiftieth birthday. Half a century. Whoop-de-do. I watched the new doctor slip the stethoscope down my father's shirt, my father scowling at me, an ashen-faced golem. This was going to be a failure. I knew it.

Dr. Bruno Cera tilted his head thoughtfully as he listened to the heartbeat. A small, lithe man with a salt-and-pepper crewcut, he spoke excellent English and gave off a relaxed, well-seasoned vibe. He also had a sense of humor—which was good. Because my father was being as nasty as possible.

"Well, Mr. Drake," said Dr. Cera in a loud voice, "I am going to take your pulse now."

"You idiot, I'm dead."

Dr. Cera laughed and held my father's wrist. "Yes, you keep saying that. Let me take your pulse anyway. . . . Well, Mr. Drake, you aren't dead yet."

My father snarled and stuck out his tongue at him. Alarmingly, it was scored with long, vertical cracks. They looked painfully deep.

"You are the one step ahead of me," said Dr. Cera, pulling out a wooden tongue depressor. "Now I can examine your throat and your mouth. Mr. Drake, say *ahhhh*."

My father bit down on the oversized popsicle stick.

"Open . . ."

My father spat it out. "Get out of here. Go on, get out."

Dr. Cera smiled cryptically and packed his things.

"Okay, we are all done here. Very nice to meet you, Mr. Drake," he said, holding out his hand, which my father ignored. "I'll be seeing you again, soon."

"You're crazy," said my father. "Get out of here!"

In the dining room, Dr. Cera and Señora Lucinda examined her journal and chart of daily medications. Her hand shook as she pointed out the dosages she had not been able to get my father to take. She felt responsible for his deterioration, I realized.

"How close to the end is he?" I asked Dr. Cera.

"Oh," he laughed, "he isn't going to die yet. He has too much fight left in him."

I felt my throat tighten up. "But he's not eating or drinking. He looks like a corpse. He's given up."

Cera's green eyes softened. "I have seen plenty of patients who have lost the will to live. Do you know what they look like? They just lie there. They don't say anything; they don't do anything. But your father—he is a fighter! I *like* that."

"Please explain that to Señora Lucinda. She needs to hear it."

He turned and began talking to her in rapid-fire Spanish. The tension drained from her face.

At this first consultation (cost: eighty U.S. dollars), Dr. Cera diagnosed my father as being severely dehydrated. He prescribed an IV drip containing dextrose, vitamins, and Risperidone, to calm him down. To treat his dementia, he recommended we bring in a colleague of his, a neurologist. He also prescribed Megace, a powerful appetite stimulator. Once my father was well hydrated, the neurologist would start treating his depression.

"Yes, he is clinically depressed," said Cera, noticing my surprise. "Perhaps for a very long time."

"He is going to get well again," he added firmly. "You will see."

In the United States, a patient needing an IV in their home would contact a "home infusion therapy provider," a company that dispatches a nurse who arrives at the patient's doorstep with all the equipment, supplies, and drugs needed. In Peru, a person undergoing any medical procedure, whether at home, in a clinic, or even at the hospital, first receives a doctor's order listing all the supplies and medications that must be bought: latex gloves, anesthetic, bandages, surgical tape, needles, et cetera, down to the littlest butterfly Band-Aid. The patient is expected to arrive at the doctor's office or operating room with all the supplies in hand. In the case of someone requiring an at-home IV, that shopping list also includes getting the whole IV contraption—the IV rack, the fluids, the drip chamber, the hose, clamps—and transporting it to the patient's room before the administering nurse arrives. It is not the most convenient way to get an IV, but it cuts out the middleman and reduces costs significantly, as we found out.

Early in the morning the day after Dr. Cera's visit, Jorge ventured to a fifty-year-old apothecary in old Lima to get the twenty-five or so items on the doctor's list. Just before I caught a cab to the university, he returned with all of the medicines, supplies, and equipment. Manuel was coming at noon to insert the catheter.

"It was all the way in butt-fuck downtown," Jorge said, wheeling the metal IV rack down the hallway. "You can't imagine. By the church of El Señor de los Milagros [Lord of the Miracles]."

I knew the place. During October the area was thronged with tens of thousands of purple-clad worshippers paying homage to an icon of Christ painted after an original that had miraculously survived two major earthquakes in 1655 and 1687. All month long, there were twenty-four-hour-long processions through the streets; booming drums, ringing bells, thick clouds of *palo santo* incense; devotees (including the president of Peru) doing penance by carrying a two-ton silver litter from one site to another. Some believers went the extra mile by hobbling around the church block on their *knees*. Thank god it was March now and not October. Otherwise, it would have taken Jorge half a day to get back home.

I returned that afternoon to find my father lying quietly in bed, a catheter taped to his left hand, a bag of fluids suspended from a rack. Seated nearby on the pullout bed was Maggy, pulling another day shift. She had dark circles under her eyes and looked uncharacteristically groggy. Holding down two jobs with no days off was obviously getting to her.

"Hola, señora," she yawned.

"How is he?" I asked in Spanish.

"Calm."

I sat down by my father's bed. The overhead light was dimmed, and Chopin was playing softly in the background. Nocturne opus 9, no. 1, in B flat minor—the tender, delicate one with ripply, circular descents on the keyboard that sound like two souls leaving and returning to each other, again and again. Breathe in, breathe out. My father's left hand lay on top of the Gators blanket, the plastic catheter affixed to his skin with white tape. His chest rose and fell in a steady rhythm. He was sound asleep.

The nostrils in his large, veined nose flared in and out, like bellows.

Even in the half-light, I could see the pink hue had returned to the hollows of his cheeks.

Like a Horse

Days 20, 21, and 22. My father did nothing but sleep. The bags of liquid dextrose, vitamins, and Risperidone, regularly replaced by Señora Lucinda and Maggy, emptied into his veins. "Tranquilo," they wrote in the journal. Maggy's angular notes were accompanied by detailed drawings of multilevel houses, corrugated metal poles extending from the roof to allow for additional stories in the future. In many parts of Peru, the more stories your house has, the higher your status.

We all kept our voices down and tiptoed through the rooms, being careful not to move suddenly and make the parquet floorboards squeak. My father, like Sleeping Beauty pricked by the needle, snoozed on.

Days 23 and 24. Dr. Cera visited and declared my father "well hydrated." The shunt was removed from his hand; Jorge returned the IV rack downtown. With Cera's okay, we started my father on the antidepressant Sertraline and jump-started his appetite with Megace.

It would take several weeks for the Megace to take effect, he said.

"Did you fire Dr. Rodríguez?" I asked Jorge.

He looked cagily at his phone. "No, I said we wouldn't be requiring his services at this time and Dr. Cera was now in charge of your father."

"So, you fired him."

"No, you don't get it. I mean, yes, I fired him, but I said it in a way that preserves his dignity."

"How did you do that?"

"There are polite ways to say things in Peruvian Spanish that keep the doors open."

"Why do we need to keep the door open?"

"You never know."

— ✦ —

Day 25. "Despacio, despacio [slower, slower]," said Señora Lucinda, laughing.

She was seated next to my father, who was sitting up in bed, spooning pureed beef stew into his mouth. The rich brown sauce dribbled from his lips as he gulped each spoonful.

I stepped in with a napkin; she didn't even have time to wipe his chin.

He stared raptly as she lifted more stew to his mouth. He opened his jaw wide, like a Galapagos tortoise.

The spoon scraped the bottom of the small white bowl. He looked at her, panicked. "More," he said hoarsely.

I tried to take the empty bowl from her.

"With your permission," she said in Spanish, keeping a firm grip on it.

We walked down the narrow hallway together, Lola following behind us, her wide otter tail thwacking against the walls. Señora Lucinda had given her a big beef bone to chew on this morning, which now made her Lola's best friend.

In the kitchen, the oniony smell of beef stew was heavenly. Señora Lucinda let Lola out on the side patio and set the small bowl in the sink. She lifted the lid of the pot, tasted the stew, and adjusted the seasoning. Then she got up on a stool and took down a big magnolia soup bowl from the shelves, one of the antiques we'd brought from Florida.

"Here, señora," she said, carefully lowering the bowl with her elegant, tapered fingers.

She ladled out three generous cupfuls, the hot steam rising up into my face. She replaced the lid of the pot and took the bowl from me. A bigger spoon? I asked. Yes, she answered.

I squeezed around her to grab a few napkins from the shelves. She seemed to have grown a half-foot taller in the last few days. Her shoulders were no longer cowed, her head held high over the nourishing dishes she was concocting for my father.

I let her go first down the hallway as I trailed behind with the soup spoon and napkins. She was dressed all in white today—white pants, white shoes, a new white hooded sweatshirt. She paused at my father's doorway, steam rising from the bowl, a triumphant look on her face.

Handel's "Hallelujah" chorus played in my head.

— ✠ —

Day 26. Maggy gave us a week's notice. I had sensed it would happen. She had been working days at the nursing home, nights at our house, and now that my father was out of the danger zone, she probably felt it was safe to let us figure things out for ourselves.

We gave her a bonus, for which she was grateful. She had a project she wanted to work on, she said.

"I bet he'll sleep through the night now," I said to Jorge. I was sitting at my desk in our bedroom, working on some notes about Qoyllur Rit'i. My filing cabinet was stuffed with research on the pilgrimage: scholarly papers about the festival's pre-Columbian roots, transcriptions of interviews I had done with pilgrims and glacier experts, geological maps of the Cusco region with the sacred pilgrimage route marked in a red line. I hadn't looked at this stuff in nearly two years, and I had pulled it out to see if I could make our journey easier and cheaper this time.

"I doubt it," said Jorge from the bed. He was in his pajamas, fooling around with his Hasselblad camera.

"Why?"

"The sundowning thing those Alzheimer's brochures talk about."

"He sleeps all the time now."

"Every week is different with him."

"I think we'll be fine with just one health aide from now on. . . . So, can we talk about Qoyllur Rit'i? I just found last year's figures. We can take a bus to Cusco, rather than a plane, and save three hundred dollars. And if we can get Paco to agree to one burro, not two, that's another hundred dollars saved."

Jorge sighed behind me.

"What's wrong?" I asked, turning around.

"Um . . . ." He looked away from the viewfinder. "I don't think it's a good idea for us to go this year."

"What?" My stomach lurched, but I tried to act calm. Being hysterical was not how to persuade Jorge to do something.

"Look, your father almost died," said Jorge. "He's still not out of the woods, and he needs looking after. Going there is going to cost almost two thousand."

"Just fourteen hundred dollars this time. Maybe less."

"Are you aware what our bills are lately? How much Maggy and Señora Lucinda eat?"

I ignored him. "We can do this. I'll find a way."

"You're so stubborn."

He picked up the Hasselblad. "The Germans offered me a gig in June. Five hundred a day. In the Callejón de Huaylas."

"Tell me you didn't say yes."

"I'm thinking about it."

The next afternoon, I heard my father yelling at Señora Lucinda. "What's wrong with this house? Why won't anybody feed me around here?" There was a crash.

I rushed down the stairs. My father had broken a magnolia plate in half.

"He just ate his second lunch," she said, apologetically, picking up the pieces.

"It's fine. Don't worry about the plate. At least he is eating."

"Yes," she said, brightening. "Your *papi* is coming back to life. You're not angry about the plate, are you?"

"No," I said, giving her a hug. "He is alive, thanks to you."

I sat in the park, stewing, while Jorge Skyped with the Germans in our bedroom.

He was coming back to life. He would probably be back to his old self soon and would only need one health aide. Now was the time to plan our next trip to Qoyllur Rit'i. I had an assignment, for god's sake.

All I needed was an expert to back me up and recalibrate my know-it-all husband.

Boxed In

Day 27. I stood in the doorway as the young neurologist ran through the battery of tests on my father. This was our third go-around with the geriatric checkup spiel, but this time was different: my father was on the mend, and surely Dr. Aguirre—a tall, calm man in his midthirties who was the head neurologist at Peru's top Alzheimer's clinic—would see that my father, as someone with mild dementia, was only going to need daytime care from here out. The neurologist came highly recommended: Not only did he work hand in hand with Dr. Cera and speak excellent, if stilted, English; he had authored peer-reviewed studies on Alzheimer's biomarkers for major medical journals and gave presentations on dementia at international conferences. I found it remarkable that a person with his qualifications made house calls, but as I was learning, this was the norm for higher-end doctors in Lima with geriatric patients.

I was holding my tongue about it, but one aspect about the doctor struck me as odd: With his pallor, his dark slicked-back hair, his dramatic eyebrows, and his low hypnotic voice, Dr. Aguirre reminded me of Dracula.

"You see here, Mr. Drake," Dr. Aguirre was saying, warming up the metal chest piece of his stethoscope between his pale hands, "I am going to listen to your heartbeat in a moment."

My father smiled innocuously. "Didn't I fire you?"

"Hah hah, no, Mr. Drake," said Dr. Aguirre, tugging on the back of his shirt. "This is the first time we meet."

"That's enough of that," growled my father as the doctor's hand went in. "Just one mo—"

My father's fist lashed out and cuffed Aguirre on the jaw, raising a red welt.

"I'll get some ice," yelled Señora Lucinda in Spanish.

Five minutes later, Dr. Aguirre sat down with me and Jorge in the dining room, looking pensive as he held the ice pack to his chin.

"So, yes, Mr. and Mrs. Vera, your father has dementia, probably Alzheimer's," he began in his low, soothing voice.

"But it is a mild case," I interrupted. "Right? In the early stages?"

He shook his head. "I cannot say he definitely has the disease without giving him an EEG, but, no, it is not mild."

I felt like a lead cape had been dropped on my shoulders.

"Your father has had the disease for a while," continued Aguirre. "In addition, he is violent. It is not that common, but 10 to 20 percent of Alzheimer's patients are this way."

"So, we hit the jackpot," I said bitterly.

"Excuse me?"

"We got lucky."

"Yes," he smiled wanly, "you got lucky."

Violento, he wrote in his notes in his unhurried script.

"He needs full bloodwork done," he continued. "Call this number. A representative from the lab will come to your house. Also, we need to get your father started on medication. I see he was on Aricept in the center of rehabilitation. We are going to put him on this." He handed me four glossy square packets that looked like oversized condoms. "Exelon, five milligrams. It comes in patch form. The nurse sticks one on his back every twenty-four hours. If he tolerates this, I will increase it to ten milligrams."

"Does this drug really work?" Jorge asked.

"It is the best thing we have. It may help him retain some of his memories and abilities. He will lead a better quality of life."

I jumped in with the question I had been dying to ask: "Do you think he will get better so he doesn't need twenty-four-hour care? Like, maybe just a nurse during the day? I mean, if this Exelon works—"

One arched eyebrow went up. "Mrs. Vera, your father is very sick. This disease only gets worse, not better." He glanced at Jorge, who was listening intently. "You should consider getting two home aides and having them alternate days. For as long as he is in your house. He cannot be trusted to be supervised by just one, I am sorry to say."

Jorge shot me a look. I ignored him. I was beat. We were beat. This was our new reality.

Day 28. I sat at the desk in our bedroom, trying not to look at the email. The UPC English department was launching a new program for professional trans-

lators starting in three weeks. Just like that. They wanted me to lead the Advanced English section and create the pronunciation program for all levels. I'd be made an assistant professor. It was a massive opportunity. It was a massive amount of work and responsibility. Why they were giving it to me was somewhat surprising but not really. I was a ham in the classroom; my old performer self enjoyed being onstage; and the students here lapped it up. More importantly, I was *una americana* with a master's degree. In the United States, my MFA didn't mean jack shit professionally, but in Peru, it made me *una experta del idiomas* (a language expert). If I took the job, I would make enough money to cover my father's rising expenses and more. Oh, what the heck.

Thank you, I would be pleased to accept, I wrote back. There. That was done.

I looked at my filing cabinet, the open top drawer stuffed with articles and maps and hand-edited transcripts. Four years of work. Thwack. An arrow through the heart.

I pulled out a photograph poking from the files. Paco and I were standing by a dirt road near Mahuayani, preparing an offering to Apu Ausangate to secure a safe pilgrimage. I was bundled in my puffy, blue down jacket, holding open a green plastic bag. Paco, in a bright-red fleece jacket and his brown felt hat with the pom-poms, scowled as he sorted through the pound of dry coca leaves; he was searching for three perfect specimens to hold, silver side up, and blow over toward the snowcapped mountain in the distance, humbly asking Apu Ausangate for protection. If the leaves were broken or held upside down, Apu Ausangate would be offended and send down lightning or let us die of the cold or bury us in an avalanche. The mountains could be as demanding and vengeful as Paco's mother.

How was I going to fit all this information on Qoyllur Rit'i into an eight hundred–word article, I wondered? It was a book. A book I no longer had the luxury of even contemplating because I had to stop galivanting around in the highlands and go back to being a teacher full-time. Jorge, at least, could travel occasionally with the Germans, who paid well. I was stuck in dusty, dirty, chaotic Lima. "Lima, the horrible," as Herman Melville had called the city.

Oh well. There was no use fighting it. It was what it was. At least my father would be properly taken care of.

— ✦ —

Just looking at the filing cabinet was so painful, I had to cover it with a hand-woven Andean textile and pretend it was a side table.

Day 29. Jorge placed the call to the health care temp agency: "Yes, we need a second *enfermera técnica*." It was like ordering a pizza. With the exception of Beijing, Lima had the most freelance home health aides in the world. They came with police reports and references, which you had to check yourself, and a résumé listing their qualifications. Some had certificates in diabetes care or dementia care. Others specialized in *cocina criolla*, traditional Peruvian cuisine. We had the cooking covered with Señora Lucinda on the days she worked; rather, we needed someone who could deal with *un anciano del Estados Unidos* who was—Jorge took a deep breath—*un poco agresivo*.

Julieta worked only three shifts before she quit. Evelyn didn't last one day. After being nearly hit twice, she came to Jorge at noon with teardrops trembling on her long, dark eyelashes. "I'm afraid," she said in a choked voice, picking at the sleeve of her worn sweater. My father's violent outbursts were too much; plus, she couldn't understand what he was saying. We fed her lunch, paid her the full day in cash, and let her go. Maggy filled in for the rest of the day. She curled up on my father's couch with a mug of sugary tea and turned on a tele-novela. When my father demanded a second dinner, she burst out laughing. That made me feel better. She knew he was on the mend, even if the new-comers were writing him off as a lost cause.

Another health aide came and quit. I was no longer remembering their names. As the doctors fine-tuned my father's medications, his violence was gradually diminishing, and soon, Dr. Cera predicted, he would no longer be aggressive—but "soon" was of little comfort to a five-foot-tall woman being chased by an elderly Sasquatch demanding more beef and potato stew.

We had to pay a finder's fee for each new *enfermera técnica*. It was starting to add up. My new salary would cover it, but that wouldn't kick in until after the start of the new semester. We could not afford any more misses.

Alma and Daisy

Day 33. Early one morning in April, I opened the front door to find a small mestiza woman in her midforties standing there, sent by the agency. She was delicate and alert, with a heart-shaped face, shiny bobbed hair, and slanted green eyes, and everything about her was tidy and compact, including her first name, Alma (Soul).

"Would you like some breakfast?" I asked in Spanish, bringing her inside.

"Oh, no thank you, señora," she said. "I prefer to wait until nine. But first, please tell me about the job."

Jorge came downstairs. "Let me get you oriented," he said, freeing me to get ready for classes.

As I learned later, Alma spent the day getting to know her new patient. Her first journal notes were telling: "He speaks English, and I didn't understand what he said. It's a shame for my patient who has depression." Depression. Even with the language barrier, she understood my father's emotional state. During the day, Alma had trimmed my father's fingernails and brought him to sit in the backyard with Lola "for the sunlight." She also recorded that my father had no regular eating schedule and wrote down her suggestions for Señora Lucinda to comment on.

I began to get a good feeling about this one.

"There's one problem," Jorge said. He was pointing to the last line of Alma's résumé. The words were in such tiny type, I hadn't noticed them. *No cocinar* (no cooking).

"Let her try," I said. "How bad can she be?"

The cloudy yellow broth lay in the bowl, flecked with dried green bits and thin noodles.

Alma tied a napkin around my father's neck. He looked down at the bowl and scowled. "What is this?"

"What does he say?" Alma asked me.

"He wants to know what the food is."

"Ah, sopa de primavera," she said, smiling nervously.

"Spring soup," I translated.

My father ladled a spoonful to his thin lips and slurped. He spat it delicately back in the spoon and stood up. "Lighter fluid," he said and shuffled back to his bed.

Alma looked down. "Excuse me, señora, I do not cook."

I brought the bowl into the kitchen and drained it into the sink. As I tossed the limp noodles into the trash can, I noticed a torn green packet: Knorr Primavera Soup Mix. Alma had bought this herself. These imported instant soups were expensive for ordinary Peruvians. That was how desperate she was to keep her job.

"The Megace is making him ravenous; we need an aide who can cook," I said to Jorge over lunch.

We were at Edo Sushi Bar, in downtown Miraflores, enjoying a plate of *tiradito*. The thin slices of fresh-cut raw corvina and tuna gleamed under the restaurant's bright lights, creamy sauces dotting the square plate. I lifted a slice and laid it on my tongue: a sweet, mild fishiness, garlic, and lime flooded my tastebuds. This was Peruvian-Japanese fusion cuisine at its best, made even tastier by the knowledge that all this seafood had been caught that morning in the Pacific. No Peruvian, not even a poor one, would think of eating day-old raw fish.

"Sure, but we can't make that the sole criteria for letting Alma go," said Jorge, downing more *tiradito* with bright-green chimichurri sauce. "You have to admit, she's great at handling him."

I thought back to what I had seen the other day: Alma standing next to my father in the bathroom with a toothbrush, imitating the motions. "You, Mister Jhon," she said, handing him the brush. She squeezed the tiniest amount of toothpaste on the bristles and turned on the faucet. Their eyes met in the mirror, and she smiled. Up went her hand, down went her hand. My father just stood there. She kept up the pantomime, no irritation in her face. Ten minutes later, I peeked in on them again. My father was finally brushing his own teeth as Alma whistled "The River Kwai March."

"Yeah, she has a real knack," I said. "Too bad the other one—"

"She's too goddamn timid," Jorge said, picking up a piece of sushi. Yellow *ají* sauce dripped from his chopsticks, onto the plate. "Especially now that he's gaining in strength again."

The drops of *ají* reminded me of Señora Lucinda's heavenly chicken dish. It had helped bring my dad back to life. Perhaps . . .

"How much would it cost us to keep Señora Lucinda on just as a cook?" I asked.

"About the same as she's making now," he said. "She would be here five days a week but wouldn't be doing twenty-four-hour shifts."

"We could get a second tech to alternate with Alma."

"If Señora Lucinda would go for it," he said, pouring some soy sauce in my tiny ceramic bowl.

I dipped a piece of *maki* into the dark saltiness. "Ask her."

First one eyebrow creased, then her eyes brightened, then she looked sideways as Jorge explained the salary. Exactly what she was making now.

"So, I would not have to sleep over," Señora Lucinda said in her low, gruff voice. "I could go home to my husband at night."

"Yes."

"And I would be the chef," she said.

"Uh, yes," said Jorge, who had originally used the word for *cook*. "You will be our head chef. For all of us."

She smiled widely, showing her small teeth, like kernels of baby corn. "Like Mamainé."

"That's her?" I said, watching the video upstairs. A hefty Afro-Peruvian woman in a red-and-white dress was dancing in an open-air kitchen, a statue of San

Martín de Porres on the shelf behind her. With the red bandana on her head and the gold hoop earrings, she looked alarmingly like Aunt Jemima.

"Ignore the outfit," said Jorge. "Mamainé is an amazing cook. I've had her food in Chincha. The best goat stew of my life."

"I can't have Señora Lucinda looking like Aunt Jemima in our house."

"It's different here. The getup means you're a good cook."

"I don't care—it's racist and embarrassing!" The vehemence in my voice sent a shudder through Jorge's shoulders.

"Señora Lucinda will be our chef," I continued in a quieter voice, "but if she wants to wear a uniform, I will get her something more dignified."

"Sure, sure," said Jorge, stopping the video. He pulled a duffel bag out of the suitcase. "You know I'm going to Ica tomorrow with ARTE. This next tech we interview better be a good fit."

"You'll interview her with me, right?"

"Of course."

"I didn't know you liked goat stew."

"My mother used to make it."

I tried to imagine my mother cooking a goat stew when I was a kid. I tried to imagine my mother shopping for goat in the Bloomfield A&P with her book of Green Stamps. All I could envision was her pulling a casserole dish of Hamburger Helper out of the oven with her singed oven mitts.

Day 35. "That one is *full pilas*," Jorge said, stacking his shirts in the duffel bag. He left for Ica tonight. He'd be gone two days.

"What's *full pilas*?"

"Full batteries—energetic, like the Energizer Bunny."

"Oh," I said, picturing the chunky, thirtysomething woman with the long dark hair and sparkling black eyes we had just interviewed. Daisy was her name. "That's a good thing, right?"

"In your father's case, I'd say so," he said, stuffing socks in a side pocket. "They seemed to like each other."

"Yeah." *Like* was an understatement. My father's face had lit up when Daisy introduced herself.

"Hello, Mister Jhon," she said in a singsong voice, all dimples and smiles. She knelt next to his chair, took his hand, and stroked it. "How are you?" When she stood up, my father stared at her ample hips, neon-pink underwear visible through her thin, white nurse's pants. Whatever demons made him strike out at people appeared to be quelled by the woman's presence.

Jorge stuffed in some pants and a sweater and dialed the nursing home by the American embassy.

"But Barbara, we should watch out for something," Jorge said, holding up one finger as he connected with the home's director. I waited five minutes as he ran through the questions: How long did Daisy work there? Was she a reliable employee? Did she come on time? Was she good with irritable old people? Did the director have any complaints?

"She checks out," Jorge said after.

"Great, so you were saying there was something we should watch out for?"

"Yeah. You caught how she was a little flirtatious, right?

I pretended I didn't. We desperately needed a second *enfermera técnica*. "She's friendly," I insisted. "So?"

Jorge rolled his eyes. "Okay—she's friendly and she's hardworking and full *pilas* and—you know what? I bet Daisy can look out for herself if your father pinches her ass."

"He wouldn't do that."

"Please, he was all over the nurses in rehab."

"You never told me that."

"There's a lot of things I didn't tell you. You had enough to worry about."

"Crap," I said. First, I had to protect the aides from my father's punches, now his lechery. It was never-ending with that man.

He zipped up the bag. "I say we go ahead and hire Daisy and let that be that."

"Okay," I said, somewhat relieved. "But let's cut off this ass-pinching business at the root. Tell her she can wear jeans if she wants to, and she should push my father's hand away the next time that happens. And if we're not there, she needs to tell us about it, not sweep it under the rug. The same goes for Alma. Explain that they won't lose their jobs for telling. We need to know about his behaviors."

"Agreed," he said, glancing at my left eye. "No more of his shit. Not under our watch."

— ✤ —

The day she started, Daisy took my father for a short walk in the backyard. Down they went to Pizarro's fig tree and back, Lola trotting beside them. Daisy chattered away in Spanish and hugged him, and my father seemed more animated than usual. "Sí, sí," I heard him say, or maybe that was my imagination.

"Try taking your father for an outing in the park," Dr. Cera suggested that night.

Land of the Dead

Day 37. I got up early to help Alma bring my father to Parque Leoncio Prado. After she applied sunscreen to his face and neck, I fitted him with a floppy khaki hat, the one he used to wear in Gainesville for painting and yardwork. He reached his trembling hand and tugged the brim so low, I doubted he could see anything. I lifted it up. He tugged it down to his nose. Oh well.

"Ready to go to the park?" I asked as Alma surreptitiously adjusted his hat from behind.

Yes, he answered after a few seconds.

Alma and I heaved him out of his chair into an upright position. It was like cranking up a mechanical doll, I thought, as he placed one foot slowly in front of the other.

After jamming his arms into his overcoat, we guided him down the hallway, through the dining room to the living room. His knuckles were white as he gripped Alma's skinny arm. His progress was slow but remarkable. Just a few weeks ago, he had been gasping on his deathbed.

Just to be safe, I decided to use a wheelchair.

I opened the double-paneled front door to the black linoleum–tiled porch. Then I rolled the wheelchair onto the tiny front lawn. It was so small, the gardener sat on his butt and cut the grass with a pair of scissors.

"Okay, let's get going," I said.

Alma guided him until his feet touched the linoleum. He froze.

"Come on, it's a beautiful day," I said.

He stared at the children and couples strolling in the park. No, he shook his head. His eyes followed worriedly as a yellow ice cream cart circled around the statue of Colonel Prado.

"Here," I said, grabbing his arm. "I'll help you down the steps. There are only four."

He wrenched his arm away. His eyes were wide open, petrified.

"What's wrong?"

"They're all dead out there," he whispered. "That's where all the dead people are."

"Dead people? You're confused. They, we—we're all alive."

"They're *dead*," he said, his voice rising in exasperation. "The ghosts. I'm sure as heck not going out there."

Señora Lucinda came into the living room. She was wearing the blue dress and thick white apron I had bought her in Surquillo Market, a low chef's cap on her head. She came closer and laid a hand on my father's wrist.

"Mister Jhon, al parque?" she asked tentatively.

"No par-kay!" he yelled. She scurried back into the kitchen.

Alma patted his hand.

I asked again and again. No. No.

A few minutes later, he backed away from the door and stomped around the perimeter of the living room. When he came close to the black linoleum, he drew back, agitated.

The linoleum was the dividing line. The one marking the land of the living from the land of the dead.

Señora Lucinda and the techs talked about it during their breaks on the patio. "Mister Jhon is seeing phantoms in the park," I heard them murmuring through the window. They didn't say it like he was crazy. They said it like he had acquired a superpower.

"The dead are all around us," said Señora Lucinda, making the sign of the cross over her ample bosom. "We must be careful." Alma nodded and crossed herself.

What was I supposed to say in response to this nonsense? I wondered. "Take him to the park or else?" This would never have been an issue in the States. I thought of calling Jorge in Ica but refrained. I should be able figure this out on my own.

The next day, Señora Lucinda and Daisy showed up wearing bracelets made of *huayruros*, black and red seeds from the Amazon to protect against evil spirits. A person who didn't know about Peruvian culture would presume they were just colorful pieces of folkloric jewelry. Sometimes my students wore them when they took exams.

Was I the only person around here not losing my mind, I wondered?

Late that night, I googled *Alzheimer's* and *hallucinations*.

I sat slumped over my computer, trying to take it all in. Hallucinations and delusions were par for the course for some dementia patients in the middle or later stages of the disease. If the hallucinations got worse, the articles said, doctors could prescribe different medicines.

There was no advice, of course, about how to handle superstitious health aides. But there were tips on how to help the patient. It was useless to argue or try to convince them that the hallucinations weren't real. It was more helpful to play along with whatever fantasy they were experiencing; contradicting the patient would only make them more agitated. If the person was seeing spiders, you could play a game to "kill" them with a hankie. If they were scared of a certain area of the house, you were supposed to lead them away to a "safe zone." The goal was to soothe and distract them, not get into an argument.

I peered out the tiny crescent window at the moonlit park below. There, like a poor man's Versailles, lay the tidy, bisecting pathways that had my father so spooked. *Okay,* I thought, tired after the long day, I would play along for my father's sake.

I got on the phone with Jorge.

"What's up?" he mumbled.

"My father says the park is full of dead people. Do you—?"

"Hmmm," he said, cutting me off. "Maybe he's entering another consciousness. Maybe now that his logical brainwaves are turned to low, he can see other worlds that coexist with ours."

I was wide awake now. "You're saying maybe there are dead people out there?"

"Maybe."

"Jorge, you can't mean that. The man is hallucinating."

"Do *you* know what's out there in the universe?" he said very gently.

I sat in the dining room, drinking my lukewarm coffee. My husband was turning all Madame Blavatsky on me. My employees were wearing *huayruros.* Even the Alzheimer's Association wanted me to kill pretend spiders with a hankie. There was no use fighting it.

When Daisy came at 8:00 a.m., I took her to the laundry room, where Alma was ironing a pair of her own pants. They both looked at me expectantly.

"We have a new activity," I told them.

That day, my father began taking daily walks in the living room. Wearing his windbreaker and a floppy hat and slathered in sunscreen.

Buried

Day 40. "Cousin, how is your father?" Chata asked in Spanish, worriedly, unwrapping a plate of beef taquitos. We were in her parents' home for Sunday *almuerzo* with about thirty family members; she had found me hiding out in the kitchen with the hors d'oeuvres.

It felt good being called *prima*. I was part of the family, she was letting me know. Me, the solitary writer, the only child with no living mother. Now I had an extended family so large, we needed a wide-angle lens for group portraits.

I explained clumsily that my father was eating again and was less aggressive. Her face relaxed.

"I'm sure he is happy being with you and Jorge," she said.

Happy. I wasn't sure how to respond. He had just crawled his way out of the Land of No-Desire and still thought the world outside was full of dead people. He snapped at everyone in the house and spat on the floor. But nobody really wanted to hear that, I was learning.

"Yes," I said.

"So, are you preparing for the trip to Qollyur Rit'i? I have been looking in my old textbooks. The syncretism of the pilgrimage is fascinating, the mix of Catholic and indigenous rites. I would so like—" She glanced at my face and stopped.

"We decided to cancel our trip this year," I said. "I got a job teaching in the new translation program at UPC. It is a great opportunity," I added, with too much enthusiasm.

"Oh, congratulations," she said, giving me a hug. When she pulled away, I could see in her eyes she knew I was lying.

"Some other time you will go," she said. "And maybe next time I will invite myself along to get in your way, okay?"

Out in the dining room, Tia Teresa had arranged a spread of barbecued meats and potato dishes and vegetable salads. As I skirted by the guacamole dip, a framed photo on the wall caught my eye. It was of Chata in her late teens, probably when she was just starting her university studies. Her shining eyes held so much hope; she had no idea that in just a few years, a civil war would break out that would make conducting fieldwork in the highlands too dangerous for any scholar, from Peru or anywhere else. She had no idea the war on Shining Path would force the country's anthropologists to take a thirty-year hiatus, that she would soon meet her future husband, become pregnant, leave the university. And she would be one of the lucky ones, unlike the seventy thousand Peruvians killed in the war.

"Maybe next time I will invite myself along to get in your way." Said in such a light, joking way, you could almost miss the buried longing. Now my longing was going underground too.

We both knew how this went. We watched as the thing we yearned for receded from our grasp, a silent scream echoing inside. The scream stayed trapped inside for years. After decades of denial, our bodies became expert at suppressing any visible betrayal of emotion: no trembling lip, no catch in the voice, no teary eye. Not even our spouse might guess what was in our hearts. Only, perhaps, a very close old friend would know how to interpret the slight slippages—a microsecond's delay in answering, a downward glance, a tiny intake of breath.

Our eyes met. There was a steady ring around her soft brown irises that seemed to say, Hush, *prima*. Hush, hush.

Day 42. I left a message for Paco with Dino, the Cusco tour operator who first took us to the Andes in 2006. Dino said my canceling with Paco was a "very fortunate coincidence." Some Dutch tourists were coming into town who wanted to book a trip to Qoyllur Rit'i in mid-June, and now Paco could be part of this expedition. I hung up, partly relieved. A big group like that would pay more than me and Jorge, only I doubted Paco would see much of it. Dino would lead the expedition himself and hire Paco as a *cargador*, a porter, to haul gear and food up the mountain. Maybe Dino would pay him fifty Peruvian *soles* a day. That was about seventeen U.S. dollars. Less than seventy dollars for four days' work in extreme conditions. When Jorge and I hired Paco for two hundred fifty dollars a day in 2008, on our second trip, he made the leap from human burro to independent tour guide.

The average Peruvian *cargador* lives around forty years, studies said.

— ✦ —

A month went by. My father turned eighty-seven. We marked it with ice cream and a card, no party, no fuss. *He's not a child,* I told myself.

I kept waiting for the Cusco number to light up on my phone, but it didn't. I knew he was thinking about me and Jorge, probably cursing us. Four hundred miles away, more than fourteen thousand feet above sea level. High in the treeless altiplano, snowcapped mountains rimming the valley, glacier-fed lakes mirroring the brilliant stars overhead. The diamond-shaped lakes that pilgrims stitched on their costumes with silver sequins that glittered under the burning sun and white moon as they danced for three days and three nights for Him.

On June 18, just before the full moon, the Pleiades reappeared in the night sky. That was when the lunar festival of Qoyllur Rit'i was held this year. I wasn't there, so I had no way of knowing if the star cluster rose large and bright, signaling the harvest would be good, or if it was faint and blurry, a bad sign for planting. I wasn't there, so I didn't hear the throbbing of the drums, the choppy wheezing of the flute players as they climbed higher and higher on the path to the dwindling glacier they could no longer stand on, it had become so unstable. I didn't hear the hushed murmurs of the faithful praying in the sanctuary to the one they believed could fix it all, the Lord of the Shining Snow Star.

Jorge didn't photograph the pilgrimage. I didn't write about it. The story became some other stringer's "local color" piece, tucked in the international section that few Americans bothered to read.

The seasons kept turning.

Qolqepunku Glacier photographed in 1935 by Peruvian artist Martín Chambi (*top, left*); and in 2006 by Mary E. Davis, who visited the site with Lonnie Thompson (*top, right*). The extent of the ice retreat over seventy-one years is shown in the composite of the two photos (*bottom*). Courtesy Martín Chambi Archives, Mary E. Davis.

CONVERSATIONS

Me and the Paleoclimatologist

How quickly is Qolqepunku Glacier melting?

I begin asking myself this question in early 2006, before our first trip to Qoyllur Rit'i in June of that year. Then I happen on the work of Dr. Lonnie Thompson, a paleoclimatologist at Ohio State University, who has been monitoring the recession of nearby Qori Kalis Glacier for decades. I email him. I tell him I'm pitching National Geographic *and* Mother Jones. *He responds encouragingly. For the next few years, he will let me pick his brain on the melting of these and other vulnerable tropical glaciers, and as I correspond with him, I am always, always, aware that in the field, Lonnie Thompson is battling his own mortality when he drills for ice cores at high altitudes. He has asthma and problems with his heart. A person like him probably should not be scaling the Himalayas and the Cordillera Vilcanota, but what can you do? There aren't many paleoclimatologists in the world, and he has to obtain as many ice cores as he can before the world's glaciers—and their climate data going back tens of thousands of years—melt into nothingness.*

"Dear Barbara," he writes on April 26, 2006, "I do not have any direct measurements of Qolqepunku Glacier. However, [I] must believe it close to that of the long-term rates of retreat that we have measured on the Qori Kalis glacier just to the south. That glacier, while retreating six meters [19.7 feet] per year in the first 15 years of measurement starting in 1963, has accelerated to 60 meters [197 feet] per year over the last 15 years. Ending in 2005, that is a 10-fold increase in the rate of ice loss."

"The glacier has probably passed the threshold," he adds.

What does that mean, I ask him by phone in August 2006, after our trip.

"You take any of these ice bodies," he says. "The ice is a threshold system. It's perfectly happy if the temperature is one degree below freezing. But as soon as you hit the freezing point, they respond very quickly to that by melting because it takes

seven and a half times less energy to melt ice than to sublimate [vaporize] ice, to go from a solid to a gas, which happens at temperatures below freezing."

That is what is happening with most of the glaciers in the world, he says. Most of them are "already doomed."

This word doomed. I hear it a lot in the 2000s from people on the ground and from scientists. But not much in U.S. publications. It is denialism, he says.

"The whole purpose of the climate denial is to muddy the waters, to make it appear that we don't know what's going on," Thompson says emphatically. "But the fact is scientists do know, and at the end of the day, the science always wins. Science is about what is, not what you would like it to be or what you would pay it to be. But is. And so, at the end of the day we will deal with this issue simply because we will have no choice."

"What we're working on more these days is . . . how do the people who live in these areas adapt to the loss of these glaciers and this water supply?" he adds. "Because glaciers are really just water towers. They collect the water during the wet season, and they release it during the dry season and during drought. That maintains water flow in the rivers below. But when they are gone, then what you're going to see is much greater range, much more water being discharged in the wet season, and many of these rivers actually drying up in the dry season. In a few decades, most of Peru's glaciers will be gone, and that will be a real problem."

We talk a bit more, and then I hang up. Peru is fucked. The country is home to 70 percent of the world's tropical glaciers, and they have lost about 40 percent of their surface area since the 1970s. Their meltwater is fueling a boom in hydropower and industrialized agriculture on Peru's desert coast—asparagus, blueberries—but communities in the Andes have been decimated by glacier lake overflows. And most of Lima's water supply comes from rivers that originate from those high glaciers. When that water is gone, how will Lima survive?

The water that comes out of the taps in people's kitchens there—it originates high in the Andes, the essence of an apu, trickling down a mountain, to join streams and rivers ending in Río Rímac, on the coast. It is such a fragile system of interdependency—and climate change is upending all of that.

Will there ever come a time when climate change arrives in Americans' own front yards, I wonder in 2006?

Little do I know what lies ahead.

4

ON EVEN KEEL

Patience

Late June. Seven columns of cards arranged precisely on the dining room table. Each column one thumb's width apart. Tops of cards neatly aligned. Face up: king of clubs, jack of spades, five of spades, nine of hearts, eight of diamonds, ace of clubs, two of diamonds.

My father's mottled hand picked up the ace and placed it in the highest row to start a stack of clubs. He flipped over the next card with a crisp snap. Queen of hearts. He pursed his thin lips and studied the options.

Daisy, seated across from him, tapped the king of clubs with an index finger. She smiled, dimples creasing her cheeks: "Aquí [here]."

He looked up, annoyed, then back at the table. The playing cards were new and glossy, bought by us several years ago on vacation in Cusco but unused until now. On the back of the cards was a photo of Machu Picchu shot on a rare, spectacularly sunny day.

My father's hand hovered over the queen of hearts, eyes flitting back and forth. Finally, he placed the queen in the top row, over the ace of clubs.

"No, no, Mister Jhon. ¡Aquí!" laughed Daisy. She pointed to the king of clubs on the left.

"No," my father said petulantly.

He counted out three cards and flipped them over. *Snap.* Two of diamonds. He straightened the cards into a neat pile. Next to him sat Señora Lucinda, sipping hot cocoa and watching intently. She had visibly relaxed since he stopped hitting people a few weeks ago. The meds were working.

He set the two of diamonds in the top row by itself. Now there were eight columns.

"Ah, no, no no, necesita . . ."—Daisy pointed to the ace—"el as. No two." She was learning more English on her own with a fake copy of Rosetta Stone we had bought her at a storefront downtown, a block from the White House. If the country's president tolerated such open piracy, what was the point in boycotting the fake merchandise?

"No, no, no," said my father, breathing hard, a drop of mucus dangling from his nostril. Señora Lucinda leaned over and wiped it away with a tissue. She folded the tissue and tucked it back in the rolled-up sleeve of her sweater, where she always carried it.

My father continued flipping over cards and putting them wherever he wanted. In the top row, he alternated piles of red and black cards; below, he lined up cards of each suit, out of order. It was reverse Solitaire, Mad Hatter style. Daisy chuckled and pointed out the correct plays, but he ignored her. In between moves, she explained to Señora Lucinda how the game was really played.

I flitted between the dining room and kitchen, preparing my breakfast. Another game began. I wasn't sure if my father had given up on the first or if he thought he won. He was bent over the cards, scowling, intent on making them do what he wanted. Señora Lucinda chimed in with suggestions, her deep voice gruff but tentative—¿Aquí?

I peeked at the table; he had reversed the rules again. The cards were all out of sequence. Nine followed three followed jack. I had never known my father to forget the rules of Solitaire. This must have been what was happening at his house all last year. His synapses misfiring, his lifetime's knowledge going dark up there. I thought of the vast library of Everyman's Classics that used to line his bookshelves, now boxed in a storage unit in Gainesville, along with his three thousand records, the music that had sustained him for decades.

An image rose to mind from my teenage years: my father lying on the sofa in the darkened living room, a drink nearby, lost in the sounds of Debussy's *La mer*. His long fingers combing the fibers of the shag rug. A suffocating sadness filled me.

But Daisy and Alma and Señora Lucinda didn't feel sad about my father's condition, I told myself. They didn't know about his books or his records or what he used to be. They just saw a funny bald *americano* named Mister Jhon messing around with a deck of cards. For hours, they giggled and laughed and *No No No'd*, and never once did they get upset or irritated. I was reminded of another name for this game: Patience. Theirs was remarkable. I would have left the table long ago.

In the end, my father swept the cards from the table and went to his room for a nap. Daisy padded after him with his meds and locked the hall door. He didn't get up until after lunch.

Daisy wrote in the book in her firm, squared-off writing: "The patient and Señora Lucinda spent the morning playing Casino. Very fun."

"My dear Señora Vera, how are you?" Jaime asked in English as I opened the front door.

"Very well."

Our landlord was paying us one of his unannounced visits. A bachelor, he lived at home with his elderly mother and two ailing dachshunds and was very religious. He and Jorge had gone to Catholic school together, run by American nuns, and when we moved to Lima, Jaime graciously offered to rent us this house. He kept the rent low but apparently believed that entitled him to critique our lives. Jorge's decision not to go to weekly Mass was "very unfortunate." Jorge's nude photos were "not for the eyes of innocents." Lola was "big and fat with dirty feet," he said, clucking his tongue at some mud on the tile floor. "Not at all like Caramelo here."

He petted the head of the aged brown dachshund tucked under his arm; the dog went everywhere with him, like a wheezing throw pillow.

"Let us see the damage, shall we?" he said.

Caramelo's flat, sloping head bobbed up and down as Jaime toured the ground-floor bathroom. Jorge had emailed him a few weeks ago asking for a new mirror. The old one had cracked when my father hit it in March. He didn't tell Jaime that.

"Yes, I will buy a new one. They are so expensive, you know," he said.

I did know. The cheap plastic ones at the Sodimac hardware store were thirteen U.S. dollars. That was undoubtedly the model we would be getting.

"Oh, by the way, how is your *papi?*"

"Fine. He is much better."

Jaime backed into the hallway and glanced toward my father's room. Caramelo's ears perked up, crisp, black origami folds. From this angle, you could see my father's veined bare feet, sticking out of his sweatpants, on top of the bed.

"Does your father have a good appetite?"

"Yes. He eats like a horse."

"Really?" said Jaime. "Thanks be to God."

A week went by. My father got up at seven o'clock each morning and by eighty-thirty was at the table, trying to remember how Solitaire went. He played all morning long, getting more and more frustrated. "No," he snapped when someone tried to help. The naps lasted long into the afternoon.

Jorge and I had a powwow. Failing at this activity wasn't doing my father any good. It was time to take Daisy up on her suggestion and try something they used at the nursing home to entertain the patients.

It was noon, and I was kneeling in the stationery section of Wong supermarket on Avenida Benavides, trying to choose a coloring book for an eighty-seven-year-old man who was color-blind. The market was named Wong after the Chinese-Peruvian man, Erasmo Wong, who founded it in the 1940s; his original store grew into Peru's largest grocery chain, and at some point, it got snapped up by Chilean investors, who kept the name. Jorge and I came here every day, it seemed—to buy food for my father, food for the aides, more food for my father and the aides, diapers, Desitin. We couldn't delegate these chores. If we sent an aide or Señora Lucinda to shop, they got excited and started buying stuff we didn't need, like Toblerone chocolate or napkins with Winnie the Pooh on them. Anyway, today's task was too complex for anyone but me to handle: My father needed a coloring book he would want to work in, one that wouldn't insult his intelligence. Simple enough. All I had to do was find a book created for Spanish-speaking toddlers that would capture the attention of a narcissistic World War II veteran with an MBA and advancing dementia.

My father had been reading David McCullough; my father had been studying the battles of the Revolutionary War, I thought as I thumbed through the coloring books. Which one to pick? "My First Day at School?" "Disney Princesses"? As I sat down to ease my aching knees, two small Andean boys crouched next to me, picked up a book on race cars, and started flipping through it excitedly. They looked about six years old, but for all I knew, they could be nine or ten; Peruvians can be short by U.S. standards, and people from the poor barrios are often stunted from malnutrition. These boys looked grimy; they probably weren't from this neighborhood. Maybe they changed buses at the busy intersection outside.

A squinty-eyed security guard in commando boots came over and shooed them off: "Go away." The boys tossed the race car book on top of some Selena Gomez lunch boxes and ran to the exit, the guard cursing them. I had seen this kind of discrimination practiced dozens of times against indigenous Peruvians, but still, it shocked me: Don't let the *cholos* dirty the pages for the white señora. (*Cholo* is a racist term used to refer to Andean people, especially poor people from rural areas.) *Cholos son choros* (*cholos* are thieves).

"They aren't bothering me," I told the guard when he came back, and he shook his head like I was a clueless foreigner who didn't know how things worked in Lima, someone who couldn't suss out the *vivos* (grifters). "No, really, it's okay," I said; the guard turned his back to me and called on his walkie talkie for a clerk to clean up the display. An employee sprayed cleaning fluid on a dirty rag and made a big show of wiping down the rack. I felt like whoever this performance was supposed to impress wasn't even in the room.

In the end, I bought the race car coloring book and one about baby animals. I splurged on a set of sixty-four Crayola crayons, which cost triple what they would in the United States. Prices for imported goods, even cheap things from China, were hiked way up in Peru.

Out on Avenida Benavides, I spied the two boys. They were lounging by a National Bank ATM and sharing a small bag of Doritos. They might have been waiting for a bus or skipping school or hoping to peek at someone's access code or trying to drum up change for a two-hour combi ride to a shantytown. I had no idea.

The next morning, before my father asked for the cards, I put the coloring books and crayons on the dining table. No Casino today, I told Daisy. I opened the baby animals book to page 1, a jungle scene of a boy playing with a tiger. The idea sank in. "Ah, los dibujos [the drawings]," she said.

She opened the lid of the Crayola box and peered at the colorful, pointed rows. The lardy smell reminded me of elementary school.

"Qué bonitos [How pretty]," she said, pulling out four crayons in green, orange, black, and white.

"Mister Jhon: Colorea [color] . . . ." She put the orange crayon in his hand and pointed to the tiger standing on its hind legs. Next to it was a shirtless boy in a sarong. The boy had no race or ethnicity—he was any color you wanted him to be: white, black, brown, or "peach," as the paper crayon wrapping read in English.

"Colorea," she insisted, pointing to the tiger.

My father looked skeptically at the page. Maybe this was too babyish for him, I thought. He hadn't held a crayon in eighty years. If they even had crayons during the Depression.

"Oranje" (or-AN-hay), Daisy said.

"Orange," I told her in English.

"Oran," she said, leaving off the last sound. English consonants can be hard for Spanish speakers.

He moved his hand to the tiger's striped foot. Tentatively, he drew a downward vertical line.

"Gooood," she said, drawing out the vowel sound.

My father's hand moved up and down. Bright-orange filled the shape.

"Más [more]," she urged, making him color all the way to the lines, leaving no white spaces.

He spent ten minutes coloring in the tiger's body and face.

"Muy bien," she said, beaming.

I came over to take a look. The colors were faint but even. From long ago, I heard the chant:

> Tyger Tyger burning bright
> In the forests of the night:
> What immortal hand or eye,
> Dare frame thy fearful symmetry?

Well, now we knew. He who wielded the crayon.

"Good job, Dad," I said out loud.

"Oh well, I still have a lot of work to do," he said, dismissively.

But I noticed he was referring to it as work.

When he stood up, he wasn't irritated.

"Where's lunch?" he asked.

The end of June. We were seeing less of our friends. No more Saturday night parties or brunches in our dining room. My actress friend, Violeta, now wanted to meet me at restaurants and cafés. The one time she came over to our house, sometime in April, my father was screaming in the back room. She didn't even put her purse down.

Our only regular guest was Jorge's brother, Henry. He was in his element in the dreary, tumultuous world of capsizing elders, having overseen his mother-in-law's care two years earlier, when her kidneys failed. Since March of this year, Henry had been regularly bringing my father Mariella's homemade organic cookies, which she sold at the weekly farmers market. "Hello, John," Henry would say, a bemused smile on his face as he rested the paper bag on my father's lunch table. Sometimes my father acknowledged him, sometimes not, but Henry kept coming to our house and smiling and leaving the cookies. Maybe he knew something about elder psychology that we didn't.

Several times, my father mentioned he didn't like the cookies Henry left, but he ate them all up. Like he did everything Señora Lucinda made—her *ají de gallina*, her lasagna, her beef stew. He was ravenous. He drank two or three tall glasses of juice at one sitting and was constantly peeing. He didn't even have to walk to the bathroom. Jorge bought him a long plastic container, called a *papagallo* in Lima, and hung it by a strap on my father's headboard. It got emptied several times a day into the toilet, the deep sound cascading off the room's tile walls and high ceiling. Way, way, way up was a hinged window you had to close with a pole. Sometimes a starling flew inside and perched up there.

A new crop of translation students enrolled at the university. The program was gaining momentum. I began teaching at UPC five days a week, from morning to late afternoon. Only a handful of tenured professors had offices, so the basement staff room became my new home away from home. My fellow teachers followed a similar daily script: a midmorning *cafecito* from the mobile snack trolley, a big *almuerzo* at a nearby restaurant with friends at 1:00 p.m., a post-lunch siesta on the sagging couches near the grading tables. I felt out of place and overwhelmed by the conviviality of it all. I didn't want to eat with my coworkers every day; I didn't want to lie prostrate on the sofas and take a nap among people I didn't know. I wanted to grab an hour of lunchtime calm before facing another three-hour teaching block.

I brought my own lunch or went to a café alone. Even as I was walking out the front gate of the university, I felt annoyed at the mass of students streaming toward me, chatting and elbowing one another, none of them aware of the right-of-way rules I had unconsciously observed in the States my whole life. "Why can't Peruvians walk on the right side of the sidewalk?" I muttered to myself, anger rising, as they jovially bumped into me and called, "¡Hola, profesora!"

Go with the flow, I told myself. *It is what it is.*

These students were way nicer and more polite than my students in Gaines-ville. I didn't have to worry about my Peruvian students getting aggressive if I gave them a bad grade or writing horrible things about me on RateMyProfessor.com.

The next day, I would step into the "salmon stream," as I called it, and—it was no use. I was pissed all over again.

Maybe, I thought, if I kept living in Peru, this would all go away, my discomfort with close crowds. Maybe I would become part of the "collective mindset," as one writer termed this Latin American value. In contrast with the United States, which values individualism above all else.

Collectivism versus individualism. I wasn't sure what side of the fence I would end up on. For now, I was discovering I was more American than I had realized.

Our translation students studied two languages, English and another of their choice: Portuguese, French, or Quechua. Barely any of our students chose the former language of the Incas. I was surprised.

"Once they graduate, they can make more money translating Portuguese or French," said Dario, a charming Belgian-Peruvian man in his early forties. He had been lured from a diplomatic post in Brussels to teach in the new translation program. He was bald and had melting brown eyes with enviably long eyelashes and was refreshingly honest. I felt instantly drawn to him.

We ate our salads together at lunch and analyzed Peruvian customs.

"But I know English-speaking journalists who cover events in the Andes who need translators who speak Quechua," I said, spearing a tomato slice.

"Barbarita, you don't understand Limeños," Dario said. "Generally, they do not value Andean traditions. Quechua is considered . . . *déclassé* compared to the other languages. Our country still oppresses indigenous people. Plus, the students here know they can make more money with Portuguese and French—interpreting for mining companies, international business meetings. You don't want all that turkey, do you?"

"Go ahead," I said as he took it all.

Quechua is *déclassé*. I began to think Dario's assessment of our students' motivations was right. Out of a new class of thirty-eight students, only five had chosen Quechua as their other language. There was talk of eliminating it next semester if more students didn't sign up.

"Don't be sad, Barbarita," Dario said to me the next day. I was back on the subject of Quechua. "Peru has many steps to go before it reclaims its place in the world and respects its heritage." He showed me photos of himself and his Peruvian lover, the two of them waving a rainbow flag at a recent gay rights parade downtown. It was one of the first such protests in Lima. "Isn't that progress?" he asked, flashing a photo of the two of them kissing on the cathedral steps.

"You guys are so cute," I said.

"You're jealous."

"No, I understand. You're my work husband. He gets to have you at home."

"But my darling, I want to put you on my *llavero* and take you with me."

"What's that?"

"Your Spanish is awful, isn't it? *Llavero*, key ring. Oh, I forgot." He dug in his attaché case. "This is my friend Nati. She's a Quechua-English translator in Cusco. Very good and—cheap? Do I say that in English? No, affordable. *Cheap* is pejorative, right?"

"I didn't ask you for the name of a translator," I said, taking the card.

"I know you will make good use of it, Barbarita. Qué bárbara. Did you know that in Spanish, your name means 'awesome'?"

My days were filled with classes and grading and lesson plans. The other hours were filled with shopping and caring for my father. It had been months since I posted on my blog. I couldn't even look at the filing cabinet with the Qoyllur Rit'i stuff in it. It was hidden in a corner, underneath that bolt of Andean fabric, but I knew it was there.

I designated the cabinet a phone charging station and put a clay vase on top, painted with white and black llamas. The llamas stared off politely into the distance as Jorge and I slept or made love or sat up late at night reading articles on "How to Manage Sundowning" and "Ten Ways to Get Someone with Dementia to Change Their Clothes."

Their Hands

Some evenings, I entered my father's room and found him and the aides sitting together on the couch, holding hands, and watching a *telenovela*.

They sat side by side in the semidarkness, their faces bathed in the blue glow of the TV. Daisy stared plaintively at the screen, her dark eyes absorbed in the unfolding drama. Her soft arm rested on a pillow propped between them, fingers intertwined with his. If Alma was watching him, she delicately patted his hand from time to time. My father looked peaceful, his face a contented blank. They could sit for hours like this, chastely, like children on a pretend date. Eventually, my father's head would nod, and he would begin to snore. He would sleep there, upright, or the aide would transition him to the bed.

As the months went by, my father became more trusting of these women's hands. They were there to guide him to the bathroom, to walk him around the perimeter of the dining room, to dab his nose when it was runny. In his former life, he would visibly stiffen when someone went to hug him; just the possibility of someone touching his comb-over made him wince. Now he sat tranquilly in the lounge chair as Daisy snipped at his balding pate with a pair of ornate barber's scissors. The newly sharpened blades flickered around his large ears. He stared out the back window at Pizarro's fig tree, regally, like he had all the time in the world. When Daisy was through, she would dab the fringe with hair gel—Ego for Men brand—and comb it neatly into place, his bald skull exposed.

She stepped back to survey her work: "Finish, Mister Jhon."

"Really?" he asked, brow furrowing as he looked up at her.

"Sí, finito."

Señora Lucinda poked her head in the room and joined us. Daisy knelt in front of my father, pinching his cheek: "Muy guapo," she said, coyly.

Did he know what *guapo* meant? I explained it was "handsome."

"Oooh," he said, obviously pleased.

"*Guapo*," Daisy repeated, pointing at him.

"Daisy, *guapo*," he said, pointing at her.

"No, soy guapa," she corrected him, emphasizing the *ah* sound of the second syllable. "Daisy, *guapa*, Jhon, *guapo*."

"Daisy, *guapa*," he repeated.

The women laugh, pleased.

Holy shit, I thought. Were they teaching him Spanish?

The women fussed over him like a doll—a cranky, messy man-doll that they could scrub and powder and primp into submission. They arranged the few hairs left on his head, trimmed his nose hairs; he stayed absolutely still as they did this. I was secretly repulsed by his peeling skin, but that issue didn't stop Señora Lucinda and Daisy. Every other day they rubbed him down, head to toe, with gobs of Nivea or Johnson's moisturizer. His hands, protruding from the elasticized cuffs of his new blue fleece pullover, gleamed with lanolin.

One evening, I caught myself feeling jealous. Nobody massaged me all over with imported creams that cost an arm and a leg.

Project Gringo Makeover kept gaining momentum; Alma got in on it too. She began pressuring me to buy cleaning products with the new scent "Bebe," which smelled like Johnson's and Johnson's Baby Powder, only ten times stronger. Bebe was the rage in Peru; even the university cleaning staff was using it to swab down the classrooms. I finally relented to Alma's polite demands. My father's room now smelled like a nursery—all softness and sweetness and talcum powder.

Señora Lucinda was in my father's room all the time now, not just cooking and straightening his room but also helping the aides bathe and change him. She had gotten over her old traumas. While she and Alma treated my father with *cariño,* affection, Daisy was teasing. "Guapo," she kept calling him, her black eyes sparkling.

One day he grabbed her ass. Jorge saw him do it.

"Oh, he's done it before," she laughed when he asked. "They all do it."

"No, Daisy. When he touches you, say no. Push his hand away."

"That is okay?"

"Yes. He is not allowed to do that. But you can still be sweet to him, if you want."

Two days later, she and Señora Lucinda were in the small courtyard where they took their lunch every day. It was a walled-in area with creeping vines and the small herb garden that Señora Lucinda had brought back to life. The two of them had bundled my father up in a fleece tracksuit and were sitting with him, one on either side, at the glass table. Daisy's chubby denim-clad leg was pressed against his.

My father grabbed her knee. She looked over at me, nodded, and put his hand back on the table.

"Mister Jhon—monkey," she said, using the English word.

My father made a mock expression of alarm. "No. Daisy monkey," he said, pointing.

Señora Lucinda snorted into her mug. She was drinking Swiss Miss cocoa. Now that I had introduced the product to the household, she went through three or four packets a day. Her upper lip was constantly perspiring.

Daisy continued: "Señora Lucinda monkey?"

There was a dramatic pause. The two women looked expectantly at my father.

He sat there solemnly, regarding Señora Lucinda. A thin smile played on his lips.

"Yes, Señora Lucinda monkey," he said.

The three of them collapsed in gales of laughter.

Wow, I thought. I had never seen my father laugh with such uncomplicated gusto in his life. Even pre-dementia, jokes usually went over his head. He didn't get the ironic ones and considered the obvious ones beneath him. These women had unlocked some forgotten chamber in his psyche.

"He is very calm and makes jokes," Daisy wrote in the notebook that night.

The Salad Bowl Incident

One evening in early July, as my father was eating dinner in his room, he looked up from his bowl of mashed potatoes. "So, how's the teaching going, Barbara?"

A jolt ran through me. He must have overheard me talking to Jorge. I graded papers in the dining room some nights.

"Um, it's okay. I'm very busy."

He smiled. "I hope they appreciate your hard work."

"Me too. Maybe they'll make me full-time faculty."

"You are a professor, right?"

"Yes, an assistant professor but not permanent. I'd like to be tenure-track."

"A tenured professor. That would be nice."

I felt woozy climbing the staircase. That was a normal conversation. My father's old self had resurfaced. *Better* than his old self—he was nice.

Over the next few days, he was lucid. He couldn't remember how to play Solitaire anymore, but he asked me and Jorge questions about our lives in Peru. Was I still doing reporting? Was Jorge working for NBC? How was his brother, Henry? It was as though he had been in a coma and needed to catch up on everything he missed. His concentration was better, his eye-hand coordination too. He filled in the figures in the coloring book with firm, decisive strokes. The sky was blue to the edges of the paper.

I was sitting at a café across from the university, having lunch by myself at the counter. I hoped no one I knew would spot my bright-green coat and feel compelled to join me. A person eating alone was to be pitied in Peru.

The café was one of hundreds of mom-and-pop restaurants in Lima that served a set *menú diario*. For the equivalent of four U.S. dollars, I could have a three-course meal with an appetizer, entrée, and dessert, plus a drink. It was cheaper and tastier than the American fast-food places that were springing up all over the city.

The waiter set a large bowl of homemade chicken soup in front of me, the hot steam wafting upward. I slipped off my knitted gloves and spooned the savory liquid into my mouth, being careful to avoid the yellow chicken's foot floating on top—a compliment from the chef. I nodded at him in the open kitchen, and he smiled, showing a row of crooked teeth.

The broth warmed me from inside, dispelling the damp winter chill. My father was returning to his old self. The rehab psychologist apparently was right. Certain cases of dementia could improve. The medications Dr. Aguirre and Dr. Cera had prescribed were working, including the antidepressant. Maybe that had been his problem all along. A lifetime of untreated depression. And me and my mother at the receiving end.

His old self was returning. The nasty edge was gone. He had asked me about my life. Could we ever talk about what had happened in the past? Would that be possible? Maybe he would do the thing I'd been wanting . . . ?

I let that hope remain unsaid.

I looked at the bowl. The only thing left was the wrinkled chicken's foot with its translucent talons. What was I going to do with this thing? This wasn't my first time trying to figure out how to avoid offending a Peruvian host who had served me the "best" animal part. Our first time at Qollyur Rit'i, Dino had Paco slaughter a sheep outside our tent and make a stew. At dinner, Paco presented me with the thigh, a jiggling four-inch-long penis attached. A great honor. Bile rose in my throat. I feigned delight and bestowed it on Paco's twelve-year-old son, who gobbled it up.

I glanced around. The chef and waiter had disappeared. I hurriedly dropped some napkins in the bowl, scooped out the chicken foot, and stuffed the napkin bundle in my coat pocket.

"How was it?" the waiter asked when he returned, eyeing the empty bowl.

"Muy rico." Compliments to the chef, I added.

A couple of hours later, I was between classes at UPC and peering over a balcony when I leaned against the lump in my pocket. Cautiously, I drew the bundle out and balanced it on the metal railing. The chicken broth had dried the napkin to a faint tea color; you could see the lumpy trident shape underneath. My lucky chicken's foot. Ordinarily there was no sun in Lima in June and July, but today the sky was lit by a pale-orange sunset, illuminating the balcony. Was that a sign?

No, I would not make a wish on it. That was preposterous.

I left the napkin behind and went back to my classroom.

On the cab ride home, I thought about a movie I once saw. A crochety old man who is estranged from his family makes friends with a troubled neighbor boy, and at the end of the movie, he contacts his daughter or son again after many years. There is a touching reunion. Tears. Hugs. Apologies. The credits roll. It's not *Home Alone* but the same idea, minus Christmas and Macaulay Culkin and the bumbling robbers.

My whole adult life, I had believed movies like that were a crock of bull. People didn't change. Mean people got meaner. People who did terrible things never said they were sorry. They just kept rolling along until one day they dropped dead, usually long after the good people had keeled over. The good people died earlier because the bad people sucked the life out of them. *C'est la vie,* as my mother used to say.

I had shut my heart to my father in my twenties. I didn't expect his love, just the occasional misshapen piece of affection gristle. I didn't expect more, and I was fine with that. That was how I had survived all these years. Plus, I had Jorge to love me.

For what it was worth, though, I had to say it. My father was laughing now. My father was asking about me. It was happening thanks to the medications and the attentive *enfermera técnicas,* who were good at their jobs and not emotionally screwed up like I was.

Change.

The next day, I was in my father's room, counting out pills with Daisy while Señora Lucinda dusted the windowsills. Lola was sniffing around the lunch table, hoping for crumbs. My father reached down to pet her.

"Where is Charlie Brown?" he asked.

I froze. Jorge had left the dog in Gainesville months ago. I didn't even have a story prepared.

Señora Lucinda stopped wiping the windowsill and stood there, watching us.

"What?" Daisy mouthed to her in Spanish.

"What happened to Charlie Brown?" he repeated, stroking Lola's head. The dog looked up at him, panting and smiling with her black lips.

"I . . . I . . . we couldn't bring him to Peru," I said. "The government doesn't allow dogs from abroad. He's living with a lady from your church now." At least that last part was true. The woman, a widow in her seventies, emailed us pictures of Charlie Brown having playdates with other dogs and going on long walks in the woods with rubber-soled booties to protect his miniscule paws. She was completely besotted.

My father stopped petting Lola. A tear rolled down his cheek. Señora Lucinda leaned in and wiped it away with the folded tissue, which she tucked back in the cuff of her sweater. I didn't meet her eyes.

"Charlie Brown is doing okay," I went on. "Mrs. Hodges from the Lutheran church is taking really good care of him. I'm sorry we couldn't bring him."

Lola nudged my father's hand with her snout.

"Where is Charlie Brown?" he asked.

"We should have him eat dinner with us," I told Jorge that afternoon. We were upstairs in the bedroom, drinking tea and eating some of Mariella's cookies, which had gone rock-hard.

"Is that such a good idea?" Jorge asked. "I thought you were grossed out by the spitting." It was true. Since my father got out of rehab, he sometimes spat his food on the floor like an angry llama. *Ptui, ptui.*

"I'm not that grossed out anymore. I just think it would be nice if we all ate together, as a family."

"Really," he said, staring at me. "You *really* think that."

"He's gotten a lot better recently. You even said so yourself. I think he might be regaining his sanity."

Jorge's eyes widened.

"It's possible. The psychologist said so."

He dunked a cookie in his tea. When he lifted it a few seconds later, half the cookie collapsed into the cup. "It's your project then," he said, stuffing the soggy rest in his mouth. "I wash my hands of it."

— ❖ —

The dark mahogany dining table, five bamboo placemats—one at the head, two on each side. Señora Lucinda busied herself around the table laying out the magnolia plates and wooden salad bowls, smiling shyly; I remembered the flicker of disapproval on her face when we told her, early on, my father wouldn't be eating with us. She had done her best to hide it then, but I had felt chastised. The cold *americana*. Her bustling expansiveness this evening made me feel less like the Grinch.

She and I had collaborated on the beef lasagna. We made the sauce from real tomatoes (her recipe), with fresh basil and garlic (my touches). Cooking from scratch was now second nature for me. An eight-ounce jar of Ragu cost the equivalent of seven U.S. dollars in Lima.

At 6:00 p.m., Alma steered my father into the dining room and lowered him onto a chair. He looked about the bright room, blinking; the aides kept the lighting in his own room subdued during the day. I jumped up to dim the lights, like I had done that first morning. Involuntarily, I pictured his balls, spinning, spinning.

"Hey, John," Jorge said.

"Hi, Dad," I said.

"Good morning," my father said, after a few beats.

Señora Lucinda and I served the lasagna and salad. We all put the napkins in our laps and started eating. Alma cut my father's food into tiny pieces. He speared a leaf of lettuce and chewed off the tip.

Please don't start spitting, I prayed, glancing at the little plaster San Martín. *Let's just have a normal meal.*

I began talking about my day at work, how the students were learning Poe's "Annabel Lee." Jorge mentioned that the new Peruvian president, a former army colonel named Humala, would be sworn in July 27, during Independence Day celebrations. My father remained silent.

"I think I get a few days off then," I said. "I have to check."

"Veintiocho de julio [July 28] is a big deal in Peru," said Jorge. "You'll probably get a whole week off."

Salad done, my father started on the lasagna. He chewed methodically and plowed through his plate, bite after bite. Not a single spit out of him. I relaxed and forgot he was there.

"Maybe that will be a good time for us to get away," Jorge said in English, switching to Spanish for Alma and Señora Lucinda. "You get three days off for the July 28 holidays," he told them. "But all three of you can't take vacation on the same dates. You need to stagger them so we have someone to cover Mister John every day. We'll put up a sign-in sheet."

"Oh, good, Señor Jorge," said Alma, taking little bites. "I want to visit my mother in Tarma."

"My daughter will be home for a few days," said Señora Lucinda. "It will be so nice to see her."

"How delicious this lasagna is, Señora Lucinda," said Alma.

"Thank you. I never made it with fresh basil before, but the señora suggested it."

Alma stepped into the kitchen to refill her water glass. Then Señora Lucinda was staring wide-eyed across the table at my father's lap.

"Holy crap!" said Jorge, leaping up.

My father was fumbling with the zipper of his pants. He was holding an empty salad bowl and trying to pee into it.

Alma rushed to grab the bowl. With Señora Lucinda's and Jorge's help, she lifted my father by the armpits and turned him away from the table to zip his pants. The two women led him back to his room.

"Never a dull moment," Jorge joked, collapsing in the chair. He dug a spatula into the tray of lasagna and scraped out some cheese and meat sauce. "Want some?"

I shook my head. My appetite was gone.

Jorge set the spatula back in the tray. "Oh well. The experiment is over."

From that evening on, the salad bowls took on a sinister aura for me. I knew they were perfectly good bowls, but when I saw their wavy, pressed-wood shape, bile rose in my throat. They were no longer bowls; they were *papagallos*. The mutation felt symbolic of something bigger, how my father's Alzheimer's was taking over our lives. The illness crept in everywhere; it was distorting, transgressive. To maintain some semblance of normalcy, Jorge and I had to set boundaries. Our own sanity was at stake.

The door stayed closed and locked during dinnertime. I wasn't Peruvian enough to integrate my father into our lives.

Three days after the salad bowl incident, my father refused to get out of bed. He threw tantrums with Daisy. He spit his pills at her. Her dimples and cajoling were to no effect. "Who are you?" he screamed, the shouts echoing down the hallway. He refused to put on any clothes and slept on top of the covers. Back to the Land of No-Desire.

Twenty-four hours later, the storm had passed. At 8:00 a.m. on a Friday, he emerged into the dining room and sat at the table. His face was pale, and his eyes looked sunken, reservoirs of exhaustion.

"Hi," I said.

He looked at me lifelessly.

Señora Lucinda laid out the coloring book and next to it some crayons: green, yellow, blue, pink. "Color, Mister Jhon."

He picked up a yellow crayon. His hand moved tentatively over the page. Thin, sketchy lines, like wisps of smoke.

Señora Lucinda watched as he tackled the top corner of the image. Round and round his hand went, planets revolving in the cosmos.

"He forgot how to put on his pants," she said matter-of-factly.

I watched, grief suffocating me, as the picture of a mother duck leading two ducklings in a field gradually took on color. Yellow streaks strayed over the mother's body onto the grass that should have been green. Then, above, an expanse of white nothingness. In the top right corner: a pink spiral, faint. Like a distant sun.

Neurons

I was starting to understand how my father's disease progressed. Right before he lost another chunk of his brain functioning, he became super-lucid. The neurons under siege rallied for a final, dazzling hoorah before they plummeted and fizzled out, like spent fireworks. Then he understood less, could do fewer things on his own.

That terrifying process must have been going on for years before his diagnosis. For long swaths of time, he had been able to hide it from everyone. His geriatrician in Gainesville even. Back in Florida, Jorge had found the briefcase my father used to bring to doctor's appointments, with his notepad inside. On the first page was a handwritten cheat sheet: His name, his age, his address, "my daughter and son-in-law live in Peru," "President Barack Obama." Answers to the doctor's questions, obviously the same ones every visit. It was heartbreaking and infuriating. Hadn't the geriatrician noticed he was reading off a yellow pad? Evidently, that didn't matter when he could bill $375 for a fifteen-minute visit and usher my vulnerable father out with flattery. "You're my star patient," my father would brag the doctor told him. The Shining Snow Star. Melting away.

For years, my already-strange father had been able to fool neighbors and acquaintances—and me and Jorge—that he was cognitively intact. But undetected, his brain grew more and more damaged, like a tropical glacier besieged by warmer air and reduced snowfall. As the glacier melted and shifted, crevasses opened up on the surface. No longer bound together by ice, rocks and boulders would dislodge and crash down the mountainside, as had started happening at Qoyllur Rit'i as early as the 1990s. Hence, the official decision in 2004 to stop pilgrims from cutting ice from the glacier, to conserve what was left. But by then it was too late. The glacier was on hospice. It could no longer regenerate itself.

That was now my father's brain: an unstable entity retreating into chaos. The chief difference: right before he took another giant step backward, he projected the mirage of being a capable, functioning person.

— ✤ —

Already, I was dreading the next time my father showed signs of improving. Facets of his former self bobbed to the surface, tantalizing me with their normalcy. It was a subtle torture I could not have imagined before Jorge and I became his caregivers.

Way back in the merry old days before my father's Alzheimer's diagnosis, I had assumed the hardest part of the disease for family members was when the afflicted person forgot their names or no longer recognized them. Sure, I knew that would be horrible whenever it happened to us. But in the meantime, I was becoming familiar with a more insidious anguish: catching glimpses of my father as he once was. Those moments disoriented *me*.

I could accept that my father had a degenerative brain disease when he was screaming that everyone in the room was dead; then it was obvious I had to take charge. It was harder to do so when he calmly looked me in the face and asked, as he still did from time to time, "Who's taking care of my house in Gainesville?" or "Did you pay the taxes for last year?" Then my necessary mental construct of him as an out-of-it Alzheimer's patient unraveled—"necessary" because it helped me make sense of his violence and grotesque behavior. If I doubted for even a minute that he had lost his mental competence, if I entertained the idea that he might be functional after all, then I was no longer sure what was true about him and what was not. My mind would keep replaying the events that had led to that moment, checking cause and effect. I would start to doubt my own judgment. My sanity began to fray.

Who was my father, I asked myself? Who had he been before? Who was he now? Which one was the real John Drake? Was there a "real" self in there, the same self that twenty years ago would have been horrified at himself for punching me in the eye or running around naked or peeing in a salad bowl? Or had that essence of my father fled, bled out into the ethers?

Lately I was reading blogs written by family caregivers for people with Alzheimer's. I had observed a common trope. Posts celebrating the brief return of a loved one's former self. A favorite song triggering the person to sing or recount a memory. A puppy's slobbery kiss bringing smiles to a catatonic spouse. See? The true self never leaves, the writers insisted. The true self is only obscured by the neurological damage wrought by the disease.

I didn't agree. Alzheimer's can turn a person into someone radically different, I was learning. Take my father. My new father was vastly different from my old father. And my new father was morphing every day into an even more extreme version of Not My Original Dad, as imperfect as he had been.

The process was like watching a building being demolished. The structure stands, then another floor collapses in on itself. But for long periods of time, everything holds. That was what his Alzheimer's brain was like.

The last three months had taken their toll on me. My father was suffering and just getting worse, and nothing could stop it. I needed a vacation. And I wasn't even the person who wiped his butt several times a day.

It was 1:45 p.m., mid-July, in the staff room. Half the professors were slumped on the couches, snoozing, sun from the skylights bleaching the cushions. I couldn't find Dario or anyone else I knew to ask.

In the copy room, I spied a bulletin board covered with flyers listing important semester dates. The only problem, the board was on the other side of the counter, and I was nearsighted. I would have to do something I dreaded: ask one of the copy guys for help.

Back in the United States, it would have just been a straightforward ask: "Can you tell me what days we have off for Independence Day?"

That was not how things were done in Lima. I would have to prostrate myself rhetorically and use the stock phrases that unlocked favors: Excuse me, a thousand pardons, can you do me a little favor (*un favorcito*)? Et cetera, et cetera.

Jorge and Dario had taught me these verbal flourishes. The first time I used them, I was shocked: They worked like a charm. Now I used them anytime

I needed something from a Peruvian in an official capacity. I tried not to feel sleazy and manipulative, but the American in me cringed.

The copy room *jefe* noticed me at the counter. I used the Ali Baba phrases and watched his stodgy face soften into a smile before he turned and checked the holiday list: One week, July 25 to 29, señorita, he said, flattering me in return by calling me "Miss." Ha. Anybody with eyeballs could see I was well past forty.

You're so nice, I told him. Very, very genteel. A thousand thank-yous.

I didn't dwell on what I had just said because I literally sounded like a supplicant at the court of Louis XIV.

Over the Threshold

It was mid-July, and the sunless sky hung over Lima like gray dryer lint, but there was one spot of brightness. My father had gotten over his fear of going outside.

It had happened the previous week while I was at work. Daisy was walking him around the house, and as they passed by the front window, she asked if he wanted to go to the park, and he said yes. She and Señora Lucinda and Jorge led him down the front steps, plopped him in the wheelchair, and pushed him across the street. He didn't throw a fit or say one word about the Land of the Dead.

Since then, he had been spending every afternoon at the park. Lola was part of the mix too. My father threw a ball to her and watched her fetch and stared at people. One of the park regulars was an eighty-year-old British man named Will, who was legally blind and went everywhere with his wife and their golden retriever. Will was the elder statesman of the park, and getting his blessing would mean my father was officially a member of the Parque Leoncio Prado inner circle.

Today Jorge, Daisy, and I wheeled my father to the center of the park for a first meeting. Will was sitting on the benches, feeding treats to his dog, Missy.

"Very nice to meet you, John," Will said in his clipped accent, reaching in the empty air to take my father's hand. His blue eyes had an opaque cast to them. The irises turned in the direction of your voice but looked right through you.

My father managed to croak out a "Very nice to meet you too." Their hands met.

"So, what brings you to Peru?" Will asked.

My father was silent for thirty seconds. I wondered if Jorge should have warned Will that my father had dementia.

"I am living in Peru with my daughter and son-in-law," my father finally said. "I have my own nurse."

"Well, that's marvelous," Will said. He rattled on about expat life and the good restaurants in the neighborhood. My father looked at him silently but attentively. Since Will couldn't see much of anything, this attention was lost on him.

Missy nudged her silky, pointed muzzle under Will's hand, demanding more treats. They had a weird relationship, these two. He gathered her fallen fur from the carpet and made sweaters out of it. I wasn't sure if this behavior was admirably sustainable or obsessive. The sweaters were honey brown and boxy in a matronly, Talbots sort of way; it was only when you got closer that you could see the fibers were all matted together, like a hairball.

After five minutes of small talk, Will petered out.

"Well, very nice to meet you, John," he said.

Back home, Daisy wrote in Spanish in the notebook: "And he is in a good mood. He made a new friend in the park."

My father hadn't mentioned Charlie Brown for a couple weeks. It was like the crying in his room never happened. Had he forgotten he ever had a poodle? Did he think Lola was Charlie Brown? Or had he resigned himself to never getting his dog back?

His brain was a deteriorating mystery wrapped in mysteries. Still, I wanted to understand. And I knew someone who could peel a few layers from that onion.

Squiggly Lines

A few nights later, Dr. Aguirre glided through the front door for an evening checkup. He had brought his portable EEG machine, and while I marveled at how affordable the service was (about eighty U.S. dollars for the visit and test), the neurologist was powering up his laptop and arranging electronic devices on the dining room table. When he pulled out the EEG cap, I got nervous.

It was a gray spandex thing studded with tiny electrodes, and I did not see how my father would ever consent to having it strapped on his head. Just the sight of those Medusa-like wires sprouting from his skull would probably make him lash out at someone. *Violento.*

"Call in your father," Aguirre said.

Alma guided my father into the room and lowered him into a chair. My father's reddened eyes widened as he glanced at the electrodes.

"We are going to make a special test on you, Mr. Drake," Aguirre said in English in his low, hypnotic voice.

Alma sat nearby, twisting her fingers.

My father looked up warily as the doctor stretched the cap over his bald skull. I waited for the right hook. Nothing.

My father blinked a few times and lifted his chin cooperatively as Aguirre tightened the strap under his wattle. The doctor swiped gel over a pair of loose electrodes dangling from the sides of the cap.

"This is not going to hurt," Aguirre continued, affixing an electrode to each cheek.

With the wired headset in place, my father looked like Alex in the brainwashing scene from *A Clockwork Orange*, minus the scary eyelid props. Weirdly, he seemed calm and relaxed—contented almost. Maybe this was familiar to him, being the object of a high-tech procedure. After all, he had spent the last seven years in Gainesville taking full advantage of his three insurance plans, having regular colonoscopies and skin biopsies and crowns replaced. There were advantages to working for the post office for thirty years, as he liked to brag.

Now that he was living in Peru, his insurance plans did not cover the medical care he received here. I didn't have the heart to tell him that. A couple weeks ago, when a lab technician came to the house to draw blood, my father asked if his insurance covered it. Yes, I lied. "Oh," he said, in a pleased but unsurprised voice. He probably thought the head scan happening now was also thanks to the bounty of the American Postal Workers Union.

"Now, close your eyes and please sit quietly," Dr. Aguirre told my father as he entered instructions on the laptop. A white graph appeared on the screen, then eight or so multicolored horizontal lines began tracing the highs and lows of his brain activity.

"When is dinner?" my father asked after thirty seconds.

"No talking," said the doctor.

"Shhh. Alma is going to get it for you," I said. He had just eaten at six o'clock, but second helpings never hurt.

I whispered to her to make a *triple*. She carried the sandwich out five minutes later as Dr. Aguirre finished the test.

"All done," Aguirre said, lifting the cap.

My father cautiously reached up to feel his skull.

"Did I pass?" he asked.

"With the flying colors," Aguirre said, smiling cryptically, as Alma led him back to his room.

The doctor leaned forward and studied the images on the screen, his slender fingers thrumming on his upper lip.

"Well, here you can see it." He pointed to a series of wavy lines. "Your father's alpha and beta waves—the ones we use in our daily thinking—they have slowed down. His theta and delta waves, on the other hand, are higher than normal for a waking adult."

"Huh?" I said.

"The theta waves occur when we daydream and use our imagination. The delta waves occur when we are in the very deep sleep."

"So . . . my father is entering another world."

"Yes. This is the pattern we see with the Alzheimer's patients. The higher brain functions decrease, and the lower ones, they take over."

I let that sink in. My father's brain was checking out of the hotel. His body, however, was still holed up in the junior suite. Ordering room service, thanks to the Megace.

I had known since February that his brain wasn't working right. Still, seeing the evidence there, in colored squiggles, gave the truth a heavy finality.

I gulped. "So, he definitely has Alzheimer's?"

"We cannot say 'definitely.' The only way to be sure completely is to take a sample from the brain, to do an autopsy. That we cannot do, for obvious reasons. But we can be fairly sure from this test and other imaging tests that your father has it."

"So what stage is he? The books say there are stages."

"Oh. That," said Aguirre. "I would say your father is in the middle stage, but he is changing rapidly."

"How long is he going to continue like this? I mean, now he still knows who we are, he's speaking, he's walking . . . ."

"We cannot know. Every patient with Alzheimer's is different. Some people stay on the same level for years, then suddenly one week, they forget how to put on their clothes or cannot feed themselves. Other patients sink slowly but steadily over some months."

I tried to absorb this information. I couldn't.

Dr. Aguirre wrote a prescription for a higher dose of Excelon and another for Memantine, for memory loss and confusion.

At the front door, he took my hand, his voice softening: "Your father is doing better. He is not so depressed. You are giving him a good quality of life here. You are doing the right thing for him."

I watched his thin silhouette disappear into the fine winter drizzle known as *garúa*. The situation with my father could last ten months. It could last ten years. He could be mobile and cognizant of who we were for all that time; he could quickly devolve and cling on for years as a vegetable. He was powerless to stop the process; we were powerless to know what would happen. Everyone in this hellish board game was trapped.

The only player who might have been smiling was the sadistic deity who conjured this hellish disease to punish humanity for doing something really horrible—I didn't know what. Neglecting to sacrifice a bunch of goats on Zeus's altar. Emitting those greenhouse gases. Killing all the dodo birds.

— ❖ —

July 28, Peruvian Independence Day, approached. Jorge and I prepared to do something we could not have imagined doing in March. We were going to leave my father alone with the aides and Señora Lucinda and take a vacation in the provinces.

The Tale of Virile Sweet and the Gold Eaters

To fully convey what happened next to me and Jorge, I need to go back in time to an empire that thrived long, long ago. To a time when there was no Lima, no Peru. To a time when living gods ruled the lands that stretched from the Pacific Ocean to the Amazon River Basin. When royal couriers ran thousands of miles along imperial stone roads—up towering mountains, through windswept valleys and verdant jungles, down to the parched desert coast—carrying messages coded in elaborate knotted threads tied around their waists. When those runners might occasionally pause in the cool shade of a eucalyptus tree and, if there were villagers loitering about, share tales of those divine rulers: of their vast storehouses of grain and gold, of their battles with their fearsome enemies, of their loyal wives, their hundreds of concubines, their many, many heirs.

Imagine yourself, dear reader, as one of those curious villagers. Imagine sitting at that young runner's feet, giving him a ladle of water and listening to the remarkable story he has carried all this way, to deposit deep inside your hearing. Wrapped in the noble language you both share: *Runa simi.* Quechua.

Once upon a time, there were two half-brothers whose father was king of the Empire of the Four Corners, which stretched from sea to sea and cradled all the lands and rivers in between. The king was descended from the Great Sun himself and, as is proper for a god, had married his own royal Sun sister and fathered many children with her, including his eldest son, Restless Fire, and his next eldest, Golden Rope. But the king loved other ladies besides, including a princess from the north, and in due time she gave him a strong son, Virile Sweet. So it was the king ruled from a mountain citadel known as the Navel of the World and looked forward to the day that Restless Fire would take his seat upon the throne.

One day, a messenger brought news that strange, bearded men had landed on the empire's northern coast. The foreigners carried thundersticks and were hungry for gold and silver. Curious, the king and his eldest, Restless Fire, traveled to the north, where they caught a terrible sickness that caused fever and vomiting and covered their bodies with pus-filled blisters. In short time, Restless Fire was dead, and the king knew he would soon follow. With his last few breaths, he announced that the two half-brothers would succeed him as joint regents; the younger, Virile Sweet, would rule in the north, and the older, Golden Rope, would govern in the south. This seemed like a good plan, but, alas, it was not to be, for just two years after their father's death, the two brothers went to war over who would succeed him as the One Great King. Thus began a bitter civil war that set brother against brother, north against south.

The War of Two Brothers raged for three years, and while Golden Rope commanded the larger, more prosperous army, Virile Sweet was aided by three skillful generals who had been loyal to his father. Finally, at a battle on the plains near the Navel of the World, Virile Sweet's forces defeated those of Golden Rope, whom he captured. To celebrate, Virile Sweet visited his northern mountain retreat in the Town of Thorns, known for its curative, mineral-rich waters.

As Virile Sweet lolled in the sacred Pool of the Emperor, he gave thanks for the victory the gods had granted him. Rumors reached him that a small party of Gold Eaters was making its way inward from the coast, but Virile Sweet was untroubled. He was guarded by eighty thousand warriors. He was the new king. He was a living god—so revered that after a single wearing, his clothes were burned, so celestial, his perfect body could not be touched by mortals. He had only to march with his men into the Navel of the World to claim his rightful throne. Why should he be worried about a few weak foreigners who foolishly wore metal pots on their heads that they didn't even use for cooking? Don't kill them, he told his aides. Let them come.

One mild spring day, Virile Sweet was roused from his bath by the news that two hundred Gold Eaters had arrived in the central plaza, just a two-hour walk from his palace. They had brought thundersticks and beasts that ran on all fours. They wanted to meet him there.

The next afternoon, after soaking in the baths and getting royally drunk, Virile Sweet was carried on a gold-and-silver litter by eighty noblemen to the plaza of the Town of Thorns. Accompanying the royal retinue were five thousand soldiers wielding knives and bolas. Loitering about the square were a handful of Gold Eaters, who kept their heads down and stared at the paving stones. Virile Sweet waited for the Gold Eaters' leader to appear and prostrate himself before a living god, but the only foreigner to present himself was a short man with a shaved head, who stepped forward with a stick and a square object covered with cloth. The man kneeled before Virile Sweet and opened the mysterious object, a box filled with thin white leaves decorated with black marks. Speaking through an interpreter, the bald man told two long, bizarre stories, one about a naked man and woman who met a python in the jungle and another about a weak god who let his favorite son be killed by commoners. The man ended by demanding that Virile Sweet swear fealty to the Gold Eaters' god, as well as to their king across the ocean. Outraged and confused, Virile Sweet pressed for details: What was their religion? Who was their king? The bald man held up the box and repeated: "Here it is, the word of God." Virile Sweet took the object and held it to his ear, but he heard nothing. "Why does it not speak to me?" he asked. He tossed it on the ground.

All hell broke loose.

Flanks of mounted Gold Eaters who had been hidden around the square galloped forward. The giant half-men/half-beasts charged at Virile Sweet and knocked his gold-and-silver litter to the ground. Balls of fire exploded from the rooftops. His royal warriors were struck through with deadly metal and bled on the stones of the plaza; those who could do so fled, but the enemy had closed many of the plaza gates, so most of his army was trapped. Terrified noblemen tried to stand between Virile Sweet and the bearded soldiers, but the Gold Eaters chopped off their hands, leaving bloody stumps. Then a man with a long black beard and flashing eyes pushed these men aside and came up to Virile Sweet, shouting "Santiago!" Black Beard and two of his soldiers grabbed Virile Sweet by the arms and legs—his divine body!—and dragged him to the periphery of the square. Gold Eaters stood guard over him while over the next two hours, five thousand of his soldiers were slaughtered, their bodies thrown in heaps around the plaza. The only blood shed by the enemy came from a deep cut on Black Beard's hand, received while defending Virile Sweet from a sword blow.

Virile Sweet was now Black Beard's prisoner in the Town of Thorns. To secure his freedom, Virile Sweet promised to fill a large room with half its height in gold and twice its height in silver. Word was sent to his subjects, and ransom tributes poured in from the Four Corners. Four months later, the

Gold Eaters melted down all the precious metals. The amount was more than enough to fill the room as Black Beard had demanded, and Virile Sweet was certain he would be freed. But the treacherous Gold Eaters did not release him. He remained a prisoner. Still, they let Virile Sweet rule the Four Corners from prison. On his orders, one of his generals traveled to the Navel of the World to kill Golden Rope, his family, and all the nobles loyal to his brother. That act upset the Gold Eaters a great deal. They called it treason; they said he was plotting against Black Beard too. They brought him in front of a scowling man who announced he was going to be burned alive. Virile Sweet was horrified; without an intact body for mummification, he could not join his divine father, the Sun, in the afterlife. To avoid eternal damnation, Virile Sweet agreed to kneel before the Gold Eaters' god and to take a new name, John.

Nine months after the Gold Eaters had ambushed his army in the plaza, the great leader once known as Virile Sweet was tied to a stake and strangled with an iron collar. Black Beard had meanwhile brought many more soldiers and running beasts to the kingdom. With them at his side, Black Beard marched on the Navel of the World, and the kingdom fell without a struggle. Within a few years, two hundred thousand of his people—half his empire—were dead of the fever that had killed Virile Sweet's father and Restless Fire.

Thus ended the three hundred–year–long Empire of the Four Corners.

The Conquest of Peru

This is the story of Atahualpa (whose name means "virile sweet," in the Quechua language) and his half-brother Huascar (*rope* or *chain*), the two sons of the eleventh Inca Huayna Capac (*young mighty one*) who fought each other for control of the Inca Empire and were defeated by Spanish conqueror Francisco Pizarro in the early sixteenth century. Cusco (*navel*) was the seat of the Tawantinsuyo (*Kingdom of Four Corners*), and Cajamarca (*Town of Thorns*, in reference to the abundant San Pedro cacti there) was the site of the royal retreat where the emperor, or Inca, regularly went to pray and replenish himself in the area's geothermal springs.

I present the story as it happened according to eyewitness accounts by Pizarro's companions, but since more than five hundred years have passed since the Massacre of Cajamarca and we have no firsthand reports by native Andean people who survived the slaughter, no one knows exactly how things happened. What we do know is that Atahualpa and Huascar had been fighting each other in a brutal civil war and that on November 16, 1532, Francisco Pizarro defeated an eighty thousand–strong Inca imperial army with 62 mounted cavalry, 108 foot soldiers, four cannons, and smallpox, and to this day, the Andean people have yet to recapture the glory of their former empire.

As Jorge and I were plotting our escape from Lima, we came across an ad for the Hotel & Spa Laguna Seca in the Baños del Inca (Baths of the Inca) region of Cajamarca. Set amid fragrant eucalyptus groves and small farms, it was a 1970s-era resort that catered to families and offered natural spa treatments and round-the-clock access to the same geothermal springs that Atahualpa had soaked in before Pizarro yanked him off his litter. There were trails to hike and thoroughbred horses to ride, but judging from the photos, what guests at Laguna Seca mainly did was eat, sleep, and get pampered. The lure of wallowing all day in warm medicinal waters and indulging in Swedish massages and "Goddess Treatments," in which I would be slathered in mud and covered head to toe with fresh rose petals, was too great to resist. We booked a last-minute package for five days.

In the Baths of the Inca

Eight in the morning. A crisp ray of sunshine slanted across the narrow gravel path, illuminating plumes of steam rising from the canal next to us. Jorge and I were wearing fluffy white bathrobes and terry cloth slippers—gifts from the hotel—as we padded toward the hotel's thermal baths, in the center of the spa complex. The route from our bedroom looped around the man-made canals that brought hot mineral water from Cajamarca's deep geothermal springs to the bathhouses of Laguna Seca. To cross the canals, we had to traverse delicate wooden bridges with bright-red railings that were vaguely Japanese. At the summit of one bridge, I leaned against the railing, closed my eyes, and inhaled: A clean, mineral scent filled my lungs—not at all like the rotten egg smell I had been expecting—and all around me was the soothing clatter of dry eucalyptus leaves blowing in the wind. I stood there, motionless, not wanting to break the spell: a moment of peace, seized out of four months of panic and despair. I held onto it for as long as I could.

"Hurry up, Barbara," Jorge urged. He wanted to be first in the bathhouse, before the other people. He was a loner, like me.

"Oh, all right."

At the entrance to the bathhouse, Jorge dug in his pocket and handed me the room card: We were parting ways here. I had booked a morning of spa treatments; he was headed to the thermal baths and Turkish sauna. In the twenty-four hours since we arrived, he had done nothing but loll around in bathing trunks and a bathrobe, and by the oozy, relaxed look on his face, it seemed the Hugh Hefner lifestyle suited him. Me—it was another story. I had been reassured several times by aides via phone that my father was okay, and I had spent hours soaking in the pools, but, still, I was a wreck. Since March, my shoulder and neck muscles had been tied up in a giant knot that made every movement, even putting a sweater over my head, a torture. Maybe the fancy treatment I had booked for today could change that.

The "Tratamiento de Diosas" was a three-hour-long bundle of procedures designed to release my so-called inner goddess. First was a "relaxation massage" given by Sonia, a fine-featured woman who commanded me to climb up on a cushioned table. The robe got whisked off, the bathing suit too. To compensate for the fact that I was butt naked, Sonia placed tiny hand towels in strategic places, removing them as she worked on different parts of my body.

So," I asked in Spanish, "where are you from?"

"Cajamarca," she said.

"In Cajamarca—your life for all time?" I asked awkwardly.

"Of course, I don't want to live anywhere else—certainly not Lima."

I lay there, thinking that one over. I couldn't blame her. "Cajamarca is sunny and green, with a perfect Andean climate and altitude, not too high, not too low," I rattled on like a tour guide. "Cusco is fascinating, but the altitude is horrible, and there are too many tourists. And Lima, it's . . ."—I wanted to say, "It's a shithole," but that didn't strike me as vocabulary befitting a goddess—"it's chaotic," I finished.

Sonia's strong hands pressed into my shoulder muscles. I opened one eye and caught a glimpse of myself in the mirror: My face was all blotchy and looked like it had aged twenty years overnight. The tendons in my neck were like steel.

Please help me, I silently prayed.

She kneaded and kneaded. The muscles gradually softened. It killed where her fingers dug in.

Before my next treatment began, I lay face up on the bed listening to piped-in New Age music. Thoughts swam up to my consciousness. Not of my father but of my mother. It had been seven years since she died of lung cancer, but her death was still painful. "Mom," I whispered, "I wish you were here. Where are you?" All these years, and I couldn't accept—what? What couldn't I accept? Her being dead? No, I knew she was gone, everybody had to leave the planet sometime. No, what my mind refused to accept were the details—two, to be exact. No. 1: Where exactly did my mother go? And No. 2: Why couldn't I contact her? Not knowing where she was made me frantic, like I had put a child on an airplane flight, and she had gotten lost in transit. Was she okay, I wondered? Did she go to the right place, wherever that was? All those galaxies and stars: Maybe her spirit got lost. Couldn't whoever was in charge allow just one phone call so I could know how she was getting on? A text message? Even a communiqué from someone else on her behalf—the cosmic equivalent of a camp counselor—to let me know how she was doing in the afterlife? That, at least, would answer my questions. But not knowing if she was Scared or At Peace or Learning to Play the Celestial Violin or Merged with a Black Hole or Reborn as a Baby in China, my mind wouldn't let go of it.

I felt these ideas and my love for her so hard, I wanted to puke. A stray thought flashed by, like a silver fish in a stream: *At least she isn't around to see your father lose his mind.* So true. She would have been fed up to here with his shenanigans. Maybe she had had a premonition of what was to come and checked out early. Still, I couldn't help wishing she were around to give me advice. She had never been fooled by anything. I could picture her in my mind's eye, sitting slump-shouldered on the stoop outside our kitchen door, puffing away on a Lark cigarette. I crouched down next to her.

"God, Mom," I whispered, "I need your help. What am I going to do about him? He's driving me crazy."

The phantom mother I had conjured up peered at me sideways through her thick, tortoise-rimmed glasses. She blinked a few times and took another drag of the cigarette; I could see the gears working as she weighed me, the situation.

She twisted and blew the smoke over her shoulder. "They don't appreciate it," she said. "That's the bottom line. I should have made that clearer to you."

"*They* who?"

"You spend your whole life trying to make them happy. It never works. You just keep giving and giving. Then you poop out."

"Like you?"

"I thought it wouldn't happen to you. I thought you were different from me," she said angrily.

"Mom, I might be different from you in some ways, but not like this. I mean, I have to take care of him. He's my father. I'm all he's got . . ."

She shook her head. "Good luck with that." Her voice was as tight as a fist.

She stubbed the cigarette out in the metallic-green ashtray she used on camping trips. All that was left now was a heap of feather-gray ashes and a broken cigarette butt blotted with coral lipstick.

I sobbed and sobbed. The tinkly music continued.

After a while, I grabbed a towel and dried my face. The nubby fabric scratched my eyelids, but I didn't care. I rubbed and rubbed. The towel smelled like bleach and eucalyptus.

The wooden door creaked open. The attendant told me to put on my robe again; we were going into the spa area. I wrapped the robe around me and stuffed the cheap foam sandals on my feet and shuffled after her. Now we were in a brighter, green-tiled shower area. I avoided looking at my face in the mirror. Again, the robe went on a hook, I lay face-down on a harder table, and another person, not Sonia, slathered my back and thighs in warm thermal mud. I was flipped over, and the procedure was repeated everywhere except for the strategic hand cloth area. The smell was earthy and metallic, like dirt dug from under a rock hidden deep in the forest. The mud settled into my skin, seeping into my pores, alive and insistent, like it was pulling something out of me. Those were the bad toxins, the assistant explained, before hosing me down. I could envision the toxins leaving me. Into the steam that floated out of the hut, over the canals, into the eucalyptus-scented air that stirred in the wind.

Now we were in Round 3 of the Goddess Treatment. I was standing in a white-tiled room, dimly lit. Along the walls were niches filled with candles. The lights flickered. In the center of the room was a large Jacuzzi filled with water, its surface strewn with flower petals, heap and heaps of them: red rose petals, pink, yellow, white, zinnia petals, orchids, a proliferation. I lowered my face to the steamy water and inhaled the delicious scents; a shiver ran through me. The robe slipped off, and I submerged myself. I turned on the Jacuzzi and let the pulsating jets smooth away any tension. A goddess. A goddess reborn.

Lying in a hammock strung between two eucalyptus trees, I wrote in a journal about my years in New York, the dark time when I was alone and fucked-up. I wrote and wrote and wrote. I scared myself. I kept going.

Fall 1984, age twenty-four

It's one in the afternoon, and I've just woken up in my dingy fourth-floor apartment in Washington Heights, on Magaw Place, after finishing another graveyard shift at *People* magazine. Blankets hang over the window rods to block the light. The only furniture is a futon bed, a folding metal chair, and a makeshift desk made of a plywood door and two metal trestles. And the rug. The orange-and-black dhurrie rug. I shouldn't have accepted it, but I did. My old weakness. Nice things.

I make a pot of coffee and settle in at the word processer. I'm chasing a prose poem, set at the American Conservatory of Music at Fontainebleau, where I was a fellowship student two summers ago. I can see the faded blue-and-gray paper flaking off the walls of the servants' quarters where we practiced, the clipped hedges and marble statues of Greco-Roman gods and goddesses visible outside through the large windows, and way off in the distance, the cypress-lined canals where my friend Anna got drunk one night with a pompous American conducting student we called "the Anti-Catharsis," who managed to impregnate her that one time, one cathartic squirt . . .

I get as far as describing the inside of my practice room, the delicate wallpaper, the *clop-clop* of dressage horses parading on the gravel paths, and then a broad heaviness settles on my forehead. It spreads laterally across my brain, numbing my thoughts, making me drowsy.

I slip off my chair, curl up on the rug, and close my eyes, exhausted.

This happens every day. The closer I am to finishing a story, the longer I lie on the rug.

I can lie here for hours, prostrate, not writing, mentally flagellating myself.

The wool carpet is itchy like straw. It was a peace offering from my father. He drove into the city one afternoon and unloaded it in the dirty alley behind my apartment building. I couldn't look him in the face. "Thanks," I mumbled. "See? Your old dad has your back," he joked. He left as soon as possible, scared by the Dominicans and their boom boxes. Now he brags to people, "I helped Barbie get set up in her apartment." *You liar,* I think. *You left me to starve that first year. I couldn't even afford subway tokens.*

At four o'clock, I get up, put on some clothes, and head out to *People.*

The next day, I try again.

— ❖ —

My mother's old therapist in Manhattan recommends I see one of his protégées, a former ballet dancer–turned–psychoanalyst who specializes in creative people with difficulties. This being New York, she's got lots of clients.

She's a Freudian. I'm okay with that.

"So, tell me," Katharine begins, settling into an armchair in her office on the Upper East Side. She has dark, luminous eyes and a warm, scratchy voice. I am wary of her; I trust her.

I start off explaining about my father.

"Go on . . ."

I no longer allow my father to photograph me at family events. All the pictures of me from this time show a mop of red hair, two pale hands covering my face.

Six months into therapy, I'm struck by a new feeling: I would like to play the oboe again, just for me, no audience, just for the pleasure of it.

Over the phone, I explain to my father that I'll be coming by train to pick up my instrument.

"That won't be possible," he says. "I sold your Lorée."

"What? When did you do that?"

"A month after you graduated."

"But it's my oboe!" I scream.

"I bought it for you," he says, all prissy voiced. "You said you weren't going to become an oboist, so there was no sense in having it lie around. I used it to pay off half your student loans. You should thank me."

No amount of crying or talking can convince him. He's closed shut, like an iron gate. I hang up, rattled to my core.

In his mind, it's simple. He poured all that money into me so I could become a professional musician, and I failed to deliver. Now he's just trying to recover his losses. The way he balances the figures, he probably thinks I owe him for being a failure.

What he doesn't understand is an instrument is not a commodity. An oboe or a guitar or a cello you've played on for years is part of your body, like a kidney or an arm.

He has amputated part of me and sold it.

Spring 1985

I'm still grieving about my lost oboe. A thirty-year-old man I've been dating off and on notices my despondency. He is a high school English teacher and a folk musician, tall and disheveled, recently divorced, with a loopy sense of humor.

"Can I come in?" He's lounging at my apartment door as I get ready to leave for the night shift at *People*. He holds out a long thin box. "For you, my dear."

I pry it open and pull out a dainty, green metal tube fitted with a black mouthpiece.

"Clarke Tin Whistle," the box reads. "For jigs and reels and all manner of Irish tunes."

"What the heck?"

He grins. "Look, if you're not going to play the oboe, you have to play something."

I buy a Celtic sessions songbook. In the mornings, before going to bed, I fill the bathtub with hot water, grab my penny whistle, and lower myself into the tub. I lean back, head resting against the old black-and-white tile walls, and practice trills and reels.

Steam rises from the soapy water as "Drowsy Maggie" and "The Redhaired Boy" drift out the small louvered window high overhead, over the grimy asphalt roofs of the Heights, out to the George Washington Bridge, across the Hudson River . . .

Afternoon after afternoon, I sit at my makeshift desk, a poster of Virginia Woolf hovering over me, battling The Tiredness, as I call it. I get partway through a poem, three-quarters, then the heaviness descends.

See? I tell myself. *You can't write. You haven't got . . .*

I'm perfectly aware of what is happening. I have paid assignments to write music articles and market analysis reports now. I can bang the articles out in a week or two. The Tiredness doesn't interfere with those. It only happens when I am writing my own stuff.

I am screwed up.

"Go on," says Katharine.

A few blocks west of Magaw Place, up in Hudson Heights, lives my friend Dylan, a pianist from a wealthy family in Westchester. He's emotionally screwed up too. He's adopted. Ever since I met him at Purchase, he has taken long naps when he's supposed to be practicing. Even simple Beethoven German dances knock him out.

One day Dylan lets himself into my apartment with the key I gave him and finds me lying on the rug.

"Don't move," he says.

He takes out his new Leica and starts shooting.

He blows the images up into twelve-by-eighteen black-and-white prints. A few weeks later, he collects some paintings and photos by our Purchase friends and hangs them in my apartment, and we have an art opening. A bad-tempered cat adopted from the Humane Society shreds toilet paper in my bedroom as an installation.

Not long after that, Dylan takes a chunk of money out of his trust fund, sells his Steinway, and opens an art gallery downtown.

Age twenty-five, twenty-six

Over the next couple years, my urge to lose consciousness when I'm writing dissipates. I fill several boxes with stories and poems and begin taking writing workshops downtown. I learn to listen to criticism without falling apart.

I take a job at a music publishing house in Port Chester, New York, editing an educational magazine for general music classrooms. I live in the White Plains YWCA for a year, saving up money to buy a car. I don't tell anyone at my workplace I live at the Y. Some of the other Y residents have probation officers.

One freezing winter night, I let a woman from ad sales drive me home. She's chatty, newly married, born and raised in Westchester County.

"My parents bought me and Joel a condo," she says, turning onto the main drag in White Plains. I let her drive right past the Y. "You should come visit us one day. I'd love for you to see our condo."

I count the number of times she says *condo*. It keeps me from thinking too hard about this new piece of information: Some people my age have parents who buy them a place to live. They don't leave them penniless in Manhattan and sell their thirty-five-hundred-dollar oboe and give them a seventy-dollar rug in exchange. Maybe my father isn't just mean and stingy; maybe he's a Depression-era anachronism as well. Most people his age had their kids when they were in their twenties. He was close to forty when I was born. From an earlier, tougher generation. His own father took out an insurance policy on him when he was drafted into the navy. I don't know what my grandfather said when my father came back alive. That didn't make it into "Take the Dubious Road."

I ask my coworker to let me off in front of a new high-rise with a red-and-white-striped awning.

"Bye!" I yell, waving from the sidewalk. I walk slowly up the stone steps, watching as her car disappears around the corner. I pause at the top step. Through the building's revolving glass door, I can see a well-lit lobby with a shiny marble floor and wooden paneling and a round table topped by a glass vase filled with flowers. Real ones.

"Can I help you?" a deep voice says. A doorman in a long, dark coat steps out from under the awning.

"Uh, yeah. What time is it?"

He looks pointedly at the oversized watch on my wrist. "Six thirty-five."

I blush. "Hah. I forgot."

Back at the Y, I turn the key to my tiny room with the single bed and the pancake-thin carpet and the bare sink under the window. The whole place smells stale and mothball-y, but I only pay $250 a month, which means I can save up for a down payment on a car and afterward get my own place. Maybe then I can get a cat and talk to it. I think I could write in front of a cat.

"My parents bought me a condo," I say out loud, imitating her New York accent. "Cawn-do, cawn-do. Do you want to see my cawn-do? Joel and I just moved into our condo, and you've got to see it, our condo. Honest to god, it's the biggest condo in the building."

Fourteen times.

Age twenty-six

One day, a young freelance writer comes to my office in Port Chester. He sits across from my desk and begins reciting:

> Mirrored in the window
> Her lover reading Catullus
> By candlelight.

I stop breathing.

It's my poem about an ex-boyfriend in California. I sent it months ago to a literary journal and forgot about. I have a Capezio shoebox full of rejection slips.

The wind astonishes the tiny house, the writer continues, from memory, glancing to see if I am sufficiently flattered. (He is there to pitch an article idea.)

I feel like throwing up.

I do not hire the writer.

I race to a nearby bookstore on my lunch hour and buy three copies of *Purchase Poetry Review,* flipping through the pages for "Marin County '85," my first published poem since I was twelve.

I find my byline.

Scanning the lines of my poem, I am seized by a sharp, primal fear: I am going to be punished.

The feeling is huge and obliterating, like being trapped underneath a teetering slate outcrop that's about to collapse. I have done something very bad, and they're going to find out and do something awful to me.

The feeling possesses me for days. Every waking second. I take little sippy breaths through my mouth like a hummingbird so I don't break into a million pieces.

Published/punished: the two words are now conflated in my psyche.

Age twenty-seven

I write more; I publish more. *Publish/punish* lessens its grip.

The idea for a novel starts gathering in my head, but I don't make an outline. That's going too far, some part of me thinks. The idea turns into an overly long story that exceeds the word count for anything a journal will publish. It lives on a floppy disk marked "Version 12," in a box hidden in my closet in New Rochelle.

I can't write with someone in the room. I can't write with a man in the room.

One Saturday morning, while a new boyfriend is sleeping over, I slip out of bed and turn on the word processor. I want to fix a paragraph. The idea came to me in my sleep.

As I'm revising, I feel him leaning over my shoulder. Teetering slate.

"Get out!" I scream, aware that I sound like a nutjob.

"Okay, okay," he laughs. He steps back, tripping on the blanket wrapped around his waist.

"What was that about?" he asks later.

"I'm very . . . private."

"No kidding. You know, I'm not going to judge."

"It's not that. I—"

I try to think what to say. I can't.

Later, I realize: Maybe he will judge, maybe he won't—it doesn't matter. I can't take the chance.

I am a house of cards building up, up, but one shake of my wobbly plywood table, and the edifice/artifice will collapse.

— ✤ —

Age twenty-eight

Christmas vacation. Ho ho ho. I'm back in my parents' house, sleeping in the celery-green room where I lived as a teenager, having long talks with my mother, letting her spoil me with pastries and cinnamon buns. Every day, behind my father's back, she shoves a twenty-dollar bill at me, folded in her arthritic hand.

Ostensibly, I'm here for the holidays. But there's another reason I drove up from New Rochelle.

Fifteen minutes ago, my mother drove off to the bakery to buy more cheese Danish and kaiser rolls. I'm alone with my father in the living room, sipping coffee, looking around at the familiar yellow walls, the white brick fireplace. He smiles absently at me, then goes back to reading the newspaper. One of the tortured Mahler symphonies he loves is playing on the stereo.

My stomach is churning. After years of therapy, I've finally worked up the nerve.

"Dad, I want to talk to you about what happened in college."

"What?"

"Remember? My junior year, you drove to Purchase, and I told you I wanted to become a writer, and you told me I didn't have the talent?"

My father's mouth is pressed in That Line.

"I didn't say that."

"Uh, yeah, you did," I say, hands shaking. "You said, 'Face it, Barbie, you haven't got it in you. The stuff's not pouring out of you.' That really hurt me."

"I didn't say that."

Some kind of crazy pressure is building up inside of me.

"And then when I graduated and I was living in New York City, you cut off all my money after a month. I only had eighty dollars left to my name. I had to go work in a bakery and live on ramen noodles. Why? Why did you do that to me?"

He's shaking his head, face flushed, comb-over slipping to one side.

"Barbie, I'd never hurt you," he said, his eyes filled with tears. "I'd never do that to you. Never."

"You did do it! Then you sold my oboe. Why? It was mine."

"Barbie, I've always been there for you. Remember when I drove all the way to New York with the dhurrie rug? Didn't we have fun that day?"

Argggh, I scream, hurling the coffee cup at the fireplace. It explodes in little pieces, brown liquid splattering across the painted white bricks.

Later I help my mother clean up the mess.

"I'm sorry he's like that," she says, as I sweep the ceramic chips into a metal dustbin. "There is a lot your father can't face."

"I'm sick of it," I say through my tears. "He makes me feel like I'm going fucking crazy."

She squats on the hearth with some paper towels and squirts Fantastik at the stain.

"Me too," she says quietly.

"Here, let me do that," I say, grabbing the bottle out of her crooked hand.

"No," she snaps, pulling it back. She rubs harder and harder. When she's done, I lift her up. Her balance is shaky these days.

She wraps her arms around me. I'm careful not to hug too hard, she's so hunched over.

"He did say those things. He did withhold the money—right?" I whisper into her shoulder.

"Yes, he did, Barbara," she says in a clenched voice. "Don't doubt yourself."

On the drive back to New York, the storm inside me slowly subsides. A hard, compressed residue is left, like the gray slush piled on the edges of the highway.

The universe handed me two parents. One is the best mother in the world. The other is a disappointment—a good provider when I was young but emotionally about one-quarter of what a father should be. Feeling counts more than money, so that should be factored into the parent equation.

One hundred and fifty percent (mother) plus 25 percent (father) equals 175 percent. All told, not bad.

But there is no use in expecting more out of that 25-percent parent—no apology, no understanding, no real help.

I am done with that.

Something inside me—that low door in the soft crenulated place open only to him—slams shut.

Age twenty-nine.

A year later, I quit my job as a magazine editor and drive to Miami Beach.

I land in an art deco hotel on Twenty-First and Collins, two blocks from the ocean.

Fuck you, magazine deadlines, I shout at the delirium-blue sky overhead as I float on the warm, salty waves. Fuck you, New York. Fuck you, neurotic father.

I step out of the crashing waves in my bikini and face the burning sun. I close my eyes.

"Make me whole again," I pray to the spirits of this new place.

I let the notebook and pen fall to the ground and took a long nap in the hammock. When I woke up, it was around four o'clock. A gentle breeze was rattling the leaves of the eucalyptus trees overhead. Horses and chickens wandered in the green fields below the spa. I felt at peace for the first time in months.

Dr. Aguirre was right. The rehab psychologist was wrong. My father was not going to get better. I would never be able to reconnect with his former self. I would never get him to admit what had happened or apologize. I would never be able to understand why he had been such an asshole to me when I was young. Alzheimer's had robbed us of those possibilities.

Writing it out—that was the best I had. Why didn't it feel like enough?

The Ransom Room

After three days sequestered at the resort, Jorge and I ventured outside to the town of Cajamarca.

There were cheese and yogurt stores.

A mile-long street with blocks and blocks of lawyers' offices where local families sold off their parcels of land to gold-mining companies or sued the companies for contaminating the local water supply with toxic metal runoff.

Some pizzerias with the soggiest pizza I had ever tasted.

And El Cuarto del Rescate, the "Ransom Room," of Atahualpa.

It was a small, stone building, the prison cell of the last ruling Inca, about forty feet long and twenty-five feet wide. The barren outside made me not want to enter, and as I stepped over the worn stone threshold, the depressing atmosphere intensified. The musty room was constructed of Inca-carved volcanic stones, with three trapezoidal windows set about seven feet above the ground. A vaulted wooden ceiling had been erected overhead, I guess to protect the historic site. Whatever preservation efforts had been attempted, they weren't enough: the stone floor and walls were chipped and flaking, as if someone had splashed them with acid. There was absolutely nothing of interest in this empty space except a red horizontal line, painted eight feet above the ground, marking the site of the original ceiling where Atahualpa had pointed to, indicating the amount of treasure his people had to collect in 1532–33 for his release: two full rooms of silver up to the limit and one room's worth of gold. A solitary red line. That was it. The whole place was stripped clean as if the Grinch had gotten there first.

A group of tourists inside was listening to a guide drone on in Spanish and Portuguese. I could guess the story. How all these poor Inca subjects had run around the empire gathering up every gold and silver treasure—including a portable throne made of 15-karat gold—and carted it to Cajamarca, dumping it on the floor here until it reached the red line. How the accountants of Carlos I marked each installment in their fat parchment ledgers. How Atahualpa believed he had fulfilled his side of the bargain and would be freed. How the Spanish, after melting down tens of thousands of pounds of priceless artifacts, reneged on their promise, tied the Supreme Inca to a post, and after toying with the idea of burning him at the stake, strangled him with an iron collar. Disgusting sadists. I was sure the guide didn't say that part.

Outside the entrance to the phenomenally depressing tourist attraction that was the Cuarto del Rescate, I leaned against a wall and watched the sun set over the Plaza de Armas below. An Andean man in a tall straw hat was standing nearby, smoking an acrid-smelling cigar; his stout wife was hawking homemade cheeses beside him. They weren't upset about the Cuarto del Rescate, so why should I be? What could any of us do about something that had decided the history of the Western hemisphere back before people in these parts had guns and immunity to smallpox? I would have bought one of the woman's cheeses, only I knew from having tasted samples in town that they all tasted the same. Salty but otherwise flavorless, the only difference being that some of the cheeses would be soft and mushy, others lumpy and dry. I had zero expectations of the cheeses that she had laid out so nicely on her woven blanket of blue and pink and yellow diamonds. I knew better than to repeat that history.

That night, soaking in the thermal waters piped into our hotel room bath, I had an epiphany: I did not want my father's presence taking over our lives anymore. It seemed indulgent given that we had three people to help us, but there you had it.

The tub drained, leaving murky grit at the bottom. That was it, the real problem: my resentment. Not just because Jorge and I had to pay people to take care of him, no, it was the whole long history we shared. For my own health and peace of mind, I needed to let go of it.

But how?

I didn't have a clue.

Guilt

Back in Lima, Daisy and Señora Lucinda were happy to see us; "Everything is fine, your father is good," they announced as we walked in the front door. I felt relieved—and guilty for wanting to be free of him.

Inside his room, my father was sitting up in his La-Z-Boy, eating orange sherbet and watching a show on polar bears. He looked calm and relaxed.

I knelt next to him.

"Hello, Barbara," he said, smiling.

I doubted he knew Jorge and I had been gone for five days. There was no immediate past with Alzheimer's. No long-term past either, or at least the parts of it that his narcissism had always denied. Just the polar bear present and sherbet melting on the side table. It reminded me of that William Carlos Williams poem.

> so much depends
> upon
>
> a blue dessert
> bowl
>
> glazed with orange
> sherbet
>
> beside the white
> polar bears

There was a Zen to having dementia. When the meds were balanced. When the demons were calmed. But not for the caregivers. In front of all of us was a full-grown problem that could never be resolved, not in the present, past, or future.

— ✥ —

August. More students joined the translation program. They were ecstatic to be studying English. The university did not require students to show any level of English proficiency to be admitted to the translation program, so some of them came in knowing not a word, but what they lacked in knowledge, they made up in enthusiasm. In between drilling them on verb tenses and pronouns, I gave them poems to read out loud: Byron's "She Walks in Beauty," Blake's "The Tyger," Poe's "Annabel Lee," the classics. The rhythms of English ran counter to the stresses in Spanish sentences, so it was good to let my students try their tongues on these old chestnuts—the more singsong the verses, the better.

The poem that my advanced students loved best was William Ernest Henley's "Invictus," with its stumbling rhythms and tongue-twisting consonant clusters. When they finished chanting the final lines—"I am the master of my fate / I am the captain of my soul"—it was as if they had won a marathon. They leaned back in their chairs, gloating, and chatted with their neighbors about what to order for lunch: *ají de gallina, arroz con pollo, ceviche mixto, lomo saltado,* the culinary poetry of Peru.

I was making progress as a teacher of English as a Foreign Language. But every time I entered our home, I felt mired in the reality of what our lives had become: We lived in a two-story building dedicated to caring for a deteriorating elderly man. The aides and Señora Lucinda circulated around the first floor carrying pills, dirty plates, buckets of soapy water to clean his room, freshly washed clothes to replace the ones he had soiled. His angry, nasal voice echoed through the high-ceiling rooms as he argued about putting on his clothes or taking a bath. Jorge had to make several runs a week to Wong and the drugstore to keep up with his needs; "I live at those stores," he grumbled. And of course, our entertaining days were over.

My father was smack dab in the center of our lives, all the time. Was there a way around this, I wondered?

A few days later, Jorge came up with a solution: build an in-law suite on the second floor, away from the main living spaces. We would pay for it.

That wasn't as crazy as it first sounded. Building supplies and labor were cheap in Peru. People were always adding extra floors or building extensions out back. You didn't have to get permits if you were a homeowner. Really, you could do whatever you wanted. Heck, three blocks away, some rich family was building a third floor completely out of drywall on top of their house. No one said boo.

We only had to ask Jaime for permission. Jorge was sure he'd say yes.

"What did Jaime say?" I asked that evening.

"No."

"Just no?"

"He said it wouldn't be a good idea."

"But having more livable area upstairs would increase the value of his property. He has to love that. Plus, his mother is old and frail. Surely, he must empathize with us. Peruvians respect the elderly, right?"

Jorge gave me a weird look. "He wanted to know how old your father is."

"So?"

"He knew your father was living with us, but when I told him he's eighty-seven, he . . ."

"What are you saying?"

"He's very superstitious."

I stared blankly at Jorge.

"He doesn't want anybody dying in his house."

I tried to wrap my mind around this. Jaime, a devout Catholic who had dedicated his life to caring for his infirm mother, believed the Peruvian superstition that having someone die in your house is bad luck. This belief flew in the face of Christian charity and duty to *la familia,* but there you had it.

Sometime in mid- or late September, I glanced again at my father's coloring books. He was now stripping the paper off the crayons and laying them on their sides to cover broader areas of the page. Filling in the figures in the coloring books was becoming less important to his "work." He left the insides blank and traced their outlines with thick bands of blue, green, and purple, like a vibrant aural haze.

— ✦ —

There was no warning.

One damp Saturday morning, Daisy and Señora Lucinda took my father for a walk in the park. An hour later, I heard the gate creak open and close. Then my father's voice: "No! No!"

I got up from my desk to peer at the front yard below. My father was standing on the pavers, the women holding him by the arm, as he flailed about, shouting "No! I won't go in! I want to go home!"

I shoved my feet into my "earthquake shoes" (the pair I always kept by the bedroom door in case of a *terremoto*) and rushed downstairs, onto the porch. Daisy and Señora Lucinda were trying every lure they could think of: "Come on, Mister Jhon. Vamos [let's go]." "Mister Jhon, monkey." "Lonche [lunch]." "Espaghetti ahn meatbowls [spaghetti and meatballs]."

My father put his head down and backed up to the metal fence: "No! I want to go home!"

"Dad," I said, stepping onto the lawn. "Let's go inside. Señora Lucinda made you a delicious lunch. Come on. It's time to eat."

"Tell these people to let me go!" he implored. "Where am I?"

"You're in Lima. With me and Jorge. This is your home."

"No!" he screamed. "I want to go home right now! Right now!"

The ruckus was attracting passersby. A few of our neighbors had gathered on the sidewalk, including the strange woman we suspected of running a clandestine adoption ring for wealthy foreigners. This was embarrassing.

Jorge came out of the garage, wiping his hands on a rag. "Come on inside, John," he said, holding out an arm. "It's cold out here."

My father shook off Daisy and cuffed Jorge on the shoulder.

"Jesus fucking Christ," Jorge cursed.

Señora Lucinda stroked my father's arm, panic in her eyes: "Mister Jhon, *tranquílate.*" Daisy ducked back in to wipe my father's nose.

The standoff lasted three hours. The neighbors watched as the old *americano* yelled and carried on. A D'Onofrio vendor began selling ice cream to the crowd. I had never felt so helpless in my life. How could you reason with someone whose brain was being consumed by an incurable disease?

The two women stood there placidly. "Come inside, Mister Jhon." I brought them all sandwiches, which my father refused to eat. Around three-thirty, he let Daisy and Señora Lucinda lead him up the steps and into the house.

Would a home health aide in the United States have been as patient? I wondered, bolting the front door. What about the millions of family caregivers in America who did this for free in addition to holding down a job? How was anyone supposed to manage a disease like this with no real support from the U.S. health care system?

Since Cajamarca, I had been brooding over how my father was weighing us down. Now I realized: As bad as it was, we were so fortunate to be caring for him here.

"Death, You Are Nothing"

"This is it," I said, jamming the last tombstone into the grass. "Henry, are you ready?"

"Not entirely," he yelled from inside.

I stood back to admire the front of the house: orange fairy lights, plastic skeletons hung on the fence, white cobwebs draped across the front door, a freshly carved pumpkin on the porch, a bowl of wrapped candies. Ghostly screams issued from the crescent window above, thanks to an old CD we'd brought from the States. These were the basic but essential trappings of Halloween, the most important holiday of the year, as far as I was concerned.

I pulled the rubber skeleton mask over my head, retrieved my scythe, and cinched the belt of my oversized trench coat. Death as a pervy private eye. Henry was finishing up his getup as a diabolical scientist. And Jorge was—well, I didn't know what he was doing, He had rushed out of the house at four, muttering "Surquillo No. 1," the name of the stinky market nearby whose tile floors were slippery with fish scales and freshly butchered entrails.

I peered through the iron fence, scanning the street for our first victims. Since moving to Lima four years ago, Jorge and I had made it a tradition every Halloween to dress up as ghouls and put on a show in the front yard, something no one else here did. Our house on Paula Ugarriza had become a destination for the growing crowds of trick-or-treaters who came from the poor barrios to Miraflores in search of free candy. Despite having our hands full with my father, we could not disappoint them this year, even if most Limeños thought the holiday was in bad taste or outright Satanic and rarely celebrated. Only recently had wealthier parents begun allowing their children to dress up in the overpriced costumes that Wong sold in colorful "Hallow-Wong" displays, along with one-liter bags of hard fruit candies.

As an American in Lima, I was here to show them how the holiday was done. That included coaching them to say "Trick or treat" in Spanish, rather than "Halloween," when they held out their goodie bags.

"Pleased to meet you, Señor Death," a wry voice said by my ear.

I turned to find Henry in black goggles and a mad scientist wig, calmly buttoning up a white lab coat I had "borrowed" from the university.

"Likewise, Professor," I said. "I'm glad to see you're getting into the spirit."

He scoffed. "This holiday is yet another perversion out of Gringolandia. But I will do it once, to understand the American psyche—not out in this courtyard, though. Last year, the teenagers frightened my mother-in-law."

"Don't worry," I began. "it's just fun and games—quick, we have our first customers."

Henry ducked back in the house as a tiny girl in a Minnie Mouse costume approached the gate, clutching a woman's hand. She stared up at me, wide-eyed.

"Ask the skeleton nicely," whispered the woman. "Dulces, por favor . . ."

"No!" the girl shrieked, hiding behind her. A small pirate and Esmeralda joined them, too frightened to speak. "¡Ay, Dios mío!" said one of the mothers.

I tugged the mask off my head and shook out my hair. "See? I am a lady. Like your mothers." I dug around in the candy bowl as Henry waved at them through the window, where he was filling a glass fishbowl with water.

As I dropped the hard candies into the plastic pumpkins, the parents thanked me effusively. Trick-or-treaters in the States would have grumbled about not getting the "good stuff." But here people were appreciative of any candy.

By 5:30 p.m., bands of teenagers were roving the neighborhood. Almost none wore costumes; a few had on beat-up rubber masks I'd seen for sale in Chinatown. The bigger kids shoved plastic bags from Metro supermarket through the fence, shouting, "Halloween! Halloween!"

"Truco o trato [trick or deal]!" I yelled back. "La frase correcta es 'truco o trato.'"

"Halloween Halloween."

Their bags heavier, the teenagers drifted over to a neighbor's house. Five minutes later, they were back again, demanding more. I began to sweat inside the trench coat.

Around 6:00 p.m., Jorge slipped through the front gate with a butcher's package under his arm. I followed inside as he unwrapped the white paper and ceremoniously hoisted the item—a hefty cow's brain—to the rim of the fishbowl. Gently he lowered it into the water, where it sank and bobbed to the surface like a buoy. At the press of a switch, an eerie green light illuminated the bowl from below.

"Whose idea was that?" I asked, impressed.

"Henry's," Jorge said. His brother emerged from the kitchen, his lab coat streaked with dark red splatters.

"Pig's blood," said Henry cheerfully. He brandished Señora Lucinda's meat cleaver and a rusty scalpel. "Now I am ready to perform my experiments."

I thought of my father, his own shrunken brain rattling in his cranium, locked in the back room with Daisy. Poor guy. Just to be sure, I jiggled the door handle a few times.

For the next hour, I passed out candies to roving teenagers while Henry brandished his instruments over the fishbowl, laughing maniacally. He wouldn't admit it, but he liked this.

The brain sank lower, sopping up liquid. Against the blackened living room, the green water looked suspended in midair, the brain lolling ponderously to one side.

As it grew darker, the trick-or-treaters grew bolder with their insults:

"Diablo, go back to Hell!"

"Death, you are nothing. You will never get me."

"I can beat your ass, big balls. Come over here."

Squinting through the narrow eyeholes, I backed up to the tombstones and tossed the candies from three feet away. Some of the pieces missed and fell on the grass. The rubber mask was sweaty against my face.

I dropped my scythe and fled.

Inside, Jorge lay sprawled on the couch, sipping red wine. Henry had disappeared, his meat cleaver abandoned on the coffee table.

"Shit," I said, peeling off the mask. "I can't breathe."

"Had enough?" he teased.

"They're serious. It's like I'm really Death, and they're getting back at me."

"But of course, Barb!" said Henry, reappearing from the bathroom, mopping his face. "Those teenagers out there—their families are from the provinces. They've lost family to Sendero Luminoso, to Fujimori's death squads. Can you blame them?"

"You don't have to stay out there," Jorge pointed out.

"Uh-uh. I'm not quitting yet." I fitted the mask over my head again. "This is Halloween. If they want an exorcism, so be it. Maybe this is what Peru needs."

Outside, a boy in a Superman outfit and flip-flops was kicking at the fence gate. I knew this kid. He lived in the house on the corner. His parents owned a Chinese restaurant downtown. A couple of weeks ago, I'd seen him pissing in the bushes in front of his own house when his parents were off at work.

He wrinkled his face. "Give me some candy. Halloween, Halloween."

"No. I know you pee in the bushes when the maid isn't looking."

"Puta madre [motherfucker]," he said. He looked at the bowl of candy longingly. There was only a handful of pieces left. "If I promise to be good, can I have some?"

"Maybe. Yes. No more peeing in the bushes, though. In the toilet."

By nine, it was time to close shop. As I was yanking up the tombstones, a family came up to the gate: a young couple and three children, a boy and two girls. Nobody had costumes. The children carried Metro bags, the father, a small metal bucket. They peered hopefully through the slats, commenting on the display.

The mother pointed to the cloudy fishbowl, the waterlogged brain resting at the bottom like a harpooned whale.

I took off my mask. "There is no more candy. Sorry," I said in Spanish.

The boy, who looked about eight, peered up at his father. "Papa, what did the skeleton say?"

"He does not have any more candy to give us, son," said the father. He had a thick black mustache and sad eyes.

"Papa, the skeleton is a woman."

He shifted the bucket to the other hand. "I know."

They stood there admiring the decorations, as if they were reluctant to leave. I got an idea. "Would you like the brain? It's fresh. We just got it from Surquillo No. 1."

"Oh, yes." He turned excitedly to his wife as I went inside.

The water slurped from side to side as I carried the heavy bowl down the steps. I set it on the lawn, next to the pumpkin, and unlocked the front gate. The father handed me his bucket. It was old and rusted where the seams met.

"Don't you want me to put it in a plastic bag first?" I asked.

"No, thanks, directly in the bucket."

Draining the water, I tipped the fishbowl sideways and slipped the slimy mass—*plunk*—into the metal container. The brain's crenulated ridges quivered as it flopped over.

I handed the bucket to the father. His downturned eyes sparkled with happiness as his wife looked on intently. Thank you, thank you, they said over and over again.

I blew out the spluttering light in the pumpkin and headed inside.

"What are they going to do with it?" I asked Jorge later in bed.

"Fry it up with some onions. Make an omelet with it. They say brain is very nutritious."

"Señora Lucinda is taking the pumpkin tomorrow. She's going to make a stew at home."

"That's good," he yawned. "Won't go to waste." He flipped over his pillow, felt for the cool side, and flopped down again.

I closed my eyes and imagined the man and his family squeezed with a dozen or so people into a run-down *combi* bus, the bucket carefully cradled on his knees or between his feet, damp, sooty air streaming through the open window, as the driver barreled through red lights and over potholes toward some godforsaken shantytown perched in the desert.

The desert that was growing drier and drier—thanks to the loss of meltwater from the Andes. Those glaciers, receding; my father's brain, shrinking. The *apus* and his consciousness: vanishing into thin air.

And meanwhile, resourceful Peruvians were doing what they had done for millennia: saving, recycling, repurposing what was left over.

Turning a discarded decoration into a hearty breakfast for All Souls' Day.

Fall

A week or two later. Sunday. Two in the afternoon. Señora Lucinda's day off. The damp chill in the air had dissipated. Summer was not far away.

Jorge and I were upstairs in the bedroom, relaxing. I was lying next to him on the bed, rereading an old issue of the *New Yorker* for the twentieth time. Salsa music was playing on the radio. Jorge fiddled with the new digital back to his Hasselblad.

The screams echoed up the staircase: "Señor Jorge! Señor Jorge! Help me, please!"

We ran to the dining room; no one was there, into the hallway—

My father was sprawled on the tile floor, face down, arms twisted beneath him. Alma was wedged under his right shoulder, struggling to get up.

"He's collapsed. He's collapsed!"

FIELD NOTES

Dancing for a Dying Glacier

May 18, 2008, Sinakara Valley, our second Qollyur Rit'i pilgrimage

On Sunday afternoon, Jorge and I climb to the place where two years ago the glacier ended, and we find nothing but dirt and moraine. It is a dizzying shock.

"So this is what the effects of rapid climate change feel like up close," I think, staring dumbly at the raw brown dirt. "The physical world around you stops making sense."

The glacier isn't all gone, of course. It has literally fled up the mountain, lying like a panting white tongue between two black peaks. Jorge and I eyeball the distance and argue over how far the glacier has receded: I say forty feet; he puts it at sixty to eighty. We finally agree that whatever the exact distance, it will take at least another forty-five minutes to climb up to the new terminus, and Jorge says he isn't up for the hike given that he doesn't have crampons to climb the ice safely once we arrive. I agree it isn't worth the risk.

The chuncho *dancers show up and put an end to our disagreement: The glacier has receded about twelve meters, they say, pointing to the old terminus spot. Forty feet.*

From where we are standing, we can only see a few lone figures struggling up the glacier, in contrast with the hundred or so pilgrims we saw here in 2006. Not even the unsmiling members of the Catholic Brotherhood are lurking to whip people for stealing ice: the longer trek up the mountain is apparently too much for most pilgrims. A thin layer of fresh snow dusts the mountaintops, but that cannot disguise that the so-called eternal snows of Qoyllur Rit'i are vanishing.

Jorge takes more photographs, and we turn to make our way down through the rocks and dirt. As I trudge along in my muddy hiking boots, I think, This place looks like a construction site.

A line of *comparsa* dancers performs outside the Qoyllur Rit'i sanctuary, below the glacier, June 1, 2009. Photo by Jorge Vera.

Forty minutes later, we cross over a makeshift bridge by the sanctuary and trudge toward our campsite, trying to avoid the clods of horse and burro excrement on the path. It is nighttime now. The valley is a sea of tents and pilgrims huddling under blue plastic tarps. Garbage is scattered everywhere, and sour smells emanate from open-air cooking pots, but the ugliness is countered by something more potent: the exuberant faith of the pilgrims.

All around us, people are dancing. Dancing in pairs, dancing in lines. Dancing in fancy, spangled costumes and fearsome masks. The dancing will go on all night into the next day, when Jorge and I leave. Even after we climb down the mountain, exhausted, and drive back to Cusco, they will still be dancing, all that Monday night into Tuesday—tens of thousands of them.

Dancing for El Señor de Qoyllur Rit'i. Dancing for a dying glacier.

5

THE SWITCHEROO

Good Hope

Crouched in the ambulance next to my father, a white sheet over his body, his face covered with an oxygen mask. Dried blood on his right temple where he had hit the floor. An EMT monitoring his pulse on an instrument panel. Sirens wailing. Was this the end?

When the technicians brought my father out of the house on a stretcher, our neighbors had been gathered on the sidewalk, whispering to one another. *¡Ay, Dios mío!*

I held my father's chilled hand as we sped across Avenida Benavides down toward the Costa Verde boardwalk, where the Clinica Good Hope is. To our left, colorful billboards advertised produce from Vivanda supermarket: an explosion of juice squirting from a fresh-cut orange into a sparkling glass. *Jugo fresco, jugo fresco* (fresh juice), the ads boasted.

We raced over the expressway, then stopped: the road ahead was jammed.

Our driver honked and honked, to no avail. This was Lima, where nobody moves aside for an ambulance.

My father's eyes fluttered open.

"It's okay," I told him. "We're going to the hospital." *Hopefully,* I said to myself.

He clutched the white sheet tighter.

The driver cursed and hammered on the horn. No one budged. The ambulance squeezed between the two cars ahead of us, creating a third lane. Finally, the other cars moved aside.

— ✦ —

The ER doctor was holding up a scan: "See this?"

Twin hemispheres, right and left, with two scoops gouged from the center like a partially eaten melon.

"It's your father's brain."

I had known the dementia was destroying his gray matter, but seeing it on the scan was a shock. How could he think coherently with those voids in there?

Dr. Cera entered the doctor's office. We had called him on our way to the clinic, but we didn't ask him to come. He volunteered. Apparently, this was normal for a private gerontologist in Lima.

Cera and the ER doctor conferred while my father was being stitched up. There was no clear cause for what happened, they said, but my father had definitely blacked out, perhaps due to decreased blood flow to the brain or other neurological changes.

"Now this has happened once, expect it to happen more often," Cera said.

Out in the waiting room, Alma was crying into a handkerchief. Her tears dried as the doors opened and my father was rolled out in a wheelchair. He smiled benevolently at her.

"¡Mister Jhon!" she said.

She is overjoyed he's okay, I thought. *She is overjoyed he didn't die on her watch.*

Now I had heard it from the doctors' mouths. What happened today would happen again.

Jaime's Letter

About two weeks after my father's fall, we received a letter in highly formal Spanish from our landlord.

> Esteemed Jorge and Barbara:
>
> At this moment, it comes to my attention that the contract for the lease of the property at Calle Paula Ugarriza, which began on the first of December 2007, is due to terminate on the first of December 2011 and will not be renewed. You will note with great attentiveness that this property is now under the management of my cousin, who will be utilizing it in another project in the future and therefore requires it to be unoccupied. Consequently, you and your family are graciously requested to vacate the property on November 30, 2011, and to return all keys to me.
>
> Yours in utter sincerity,
> Jaime A. Velasquez Nuñez

I understood the purpose of this flowery language. We had just been delivered a Tiffany's box with a turd inside.

"Christ almighty!"

Jorge threw the phone down and glared at me. "He won't do it."

"I figured."

"Shit! This guy went to Carmelitas with me, and now he's kicking us out on the street in just one month? We're supposed to find a place for all of us by then?"

I touched his temple. A wriggly blue vein was throbbing there.

"Don't," he said.

"What are we supposed to—?"

"Let me think."

He drew back the curtains and looked out at the park. I joined him, my stomach in knots. We had just gotten everything on an even keel—my father's health, his stability, the *enfermeras técnicas*. And now . . .

I could see Jaime in my mind's eye. In his old-man sweater-vest and pleated trousers. The dog. The dog with back problems that needed to be carried everywhere. Poor thing. I shouldn't be mad at the dog. I couldn't help picturing it. Its sloping head, the pert little ears, the wheezing.

"I have a job with the Germans in three weeks!" Jorge exploded. He was pacing by the mantel. "Up in Chaclacaya. Four hundred a day. Maybe I can get them to postpone the shoot."

"What do we do now?"

"Do? We look at houses, that's what we do. I'll do it while you're at work." He smiled at me. "It shouldn't be so hard, right?"

"Nada, Barbara."

He threw down the folded newspaper with the houses-for-let ads he had been tracking for days. Three or four properties were circled in blue pen, large Xs crossed through them. I peeked at the figures. They were twice what we were paying.

"What about this one in Barranco?" I said, pointing to an ad at the bottom. "Three bedrooms, three baths."

"That's not a house. It's an apartment. And it's tiny."

Out in the kitchen, he grabbed a Cusceño beer from the fridge and popped it open. He only drank beer when he was stressed. He plopped back down on the couch, took a sip, and sighed.

"Look, the bottom line is, nobody is renting houses for nine hundred a month anymore. They were four years ago; now everything is changing. The city is being built up; they're knocking down all the old houses to make high-rises." He sighed again. "I hate to say it: Unless a miracle happens, we aren't going to be able to find a house big enough for all of us that we can afford."

I stared at the ads, my eyes tearing up. "Why can't Jaime let us stay here? Why is he letting his cousin have the house?"

"Barbara, *cousin* is code for he's selling it to a developer. They'll knock it down and combine it with the land they just sold next door. Then they can build a bigger apartment building. Everybody gets rich."

"Oh, so that's what's going on," I said, thinking about the dodgy neighbor with the adoption business next store. Since she'd signed the deal with the developer, there had been no more Andean babies arriving in the night, no more Americans buzzing her front doorbell.

"So, what do we do now?" I asked.

"There's only one thing. A *casa de reposo.*"

The Search

Rest homes, or *casas de reposo,* as they are known in Spanish, did not have a good reputation among most Peruvians in 2011. They were places where "unfeeling" or "mean" children put their ailing elders; good sons and daughters were supposed to take care of *mami* and *papi* at home, often right up until the end. People with money hired *enfermeras técnicas.* Poor people took care of their parents themselves. By deciding to outsource my father's care, we were bucking Latin American tradition, although it was true that in recent years rest homes had begun to pop up in upscale Lima neighborhoods.

I didn't want to say it out loud. The idea of him not living with us anymore was freeing. But I was unsure we could find a place decent or affordable enough, one that wouldn't send my father plunging back into the Land of the Dead. It had to feel right. Did I have that luxury now that Jaime had upended our master plan?

With just three weeks until his deadline, the hunt was on.

Three blocks south of us on Calle Paula Ugarriza stood a converted one-family home. Bright-yellow daisies grew in the dirt outside, but everything else looked run-down: the crumbling plaster around the windows, the dusty cracked paint on the walls.

Jorge pressed the metal buzzer. A harried-looking woman poked her head out of a window upstairs and yelled down, "What?"

Afterward, I couldn't remember the tour. The scene in the communal living room obliterated everything. Eight or nine elderly people—women mostly—sitting in overstuffed chairs, clustered around an enormous Magnavox TV. The armchairs were covered with stained chintz fabric. Tied around each oldster's waist: a grimy rope knotted to the back of the chair.

"So they won't run away," explained the proprietor.

—❖—

Then there was the extremely Catholic place out in Surco. It was next door to a well-known medical clinic—a plus—with real RNs and hygienic-looking hallways and tidy common rooms. The freshly waxed linoleum floors shone like polished marble. But the residents' private rooms looked and smelled like hospital rooms, with metal beds and metal side tables and waste bins that opened with a squeak when you pressed the foot pedal and smelled pungently of bleach. The crosses nailed over the beds were made of real wood, though.

The big draw in that place were the nun-nurses. On the afternoon we visited, they were everywhere. Ditto the priests. If you died, these people were all on hand to make sure you didn't miss out on the last rites, even if it was three in the morning or in the middle of an earthquake. A true believer could really take care of final business there. In between the day you arrived and the day you kicked the bucket, you could count rosary beads and have confession and contemplate the life-size portraits of cardinals and the pope in the hallways. Nobody was smiling in those portraits; their holy eyes bored into you like the gaze of Inquisitors. The setup had all the cheer of a morgue. It was designed to comfort a devout Catholic, but that had nothing to do with my Masonic Lutheran father. He would tip back into depression and madness if we put him there.

Someplace homey, I decided. That was what he needed.

Two weeks before our ejection date, Jorge and I visited an Episcopal nursing home in Miraflores. It was housed in a turreted, two-story building by the Malecón (Boardwalk), with a little rose garden outside and an interior garden as well. The owner seemed nice enough. The rooms were only US$1,000 a month, board included, and there was an opening.

I peeked in the vacant room; it had a green linoleum floor circa 1940 and a chrome hospital bed. Well, I thought. Maybe we could bring my father's own furniture and cozy the place up. But would he like it here?

"Who would be looking after him?" I asked.

No, my father would not have a private nurse, said the owner. The same health aides took care of everyone. They never had problems.

Jorge glanced at me quizzically; we were on the same wavelength. My violence-prone, difficult father left on his own for much of the day? If this woman was claiming the residents "never" gave the staff problems, there was one explanation: They were probably overly sedated.

I stepped into the lobby to inspect the front door—just a simple lock and a deadbolt. I could imagine my father figuring out the mechanism and slipping out one night, like he did at the Gainesville rehab. And if he wanted to communicate with the staff, how could he do that?

The owner smiled. "Why, yes, we have some English speakers," she said, leading us to the interior garden.

A thin, worried-looking woman stood next to a potted rubber plant, intently pinching a leaf. She didn't look up when we greeted her.

"¡Hola, Señora Hillman!" we said again, louder. Then, "Hello, Mrs. Hillman!" She lifted her head, one dark eyebrow raised.

"We're Barbara and Jorge," I said. "From the United States. Florida."

"Florida," she said, brightening. "My sister lived in Florida," she said in English. "There was a Meyer lemon tree out back. We made jam with them."

"Where are you from?" Jorge asked. "Mrs. Hillman? Mrs. Hillman?"

She was staring off at some middle point. Her eyes wandered back to Jorge. "Hartford," she said in a faint voice. "My husband is retired from International Paper. We live in Peru now."

She looked down and went back to pinching the rubber tree.

"Her husband has been dead for seven years," the owner told us in Spanish in her office. Elaine Hillman had met her husband, a Limeño, somewhere in the United States, and the two of them had retired to Miraflores in the mid-1990s. They had a large apartment on the Malecón, overlooking the ocean. In 2004, her husband died of a heart attack. Mrs. Hillman lived alone for several years and began fighting with her Peruvian relatives. Soon there was only one family member, a nephew, who visited her, and then even he stopped. No one knew when the dementia started.

A few years ago, Mrs. Hillman started leaving her apartment and meandering up and down the Malecón for hours on end. She stopped eating and taking care of herself. Finally, a person who had known her from the Anglican-

Episcopal church in Miraflores brought her to the nursing home. Now the church looked after her. Once she had been the church's biggest donor.

I gathered up the brochure and the price quote and shook the owner's hand. "We'll get back to you," we said. The wrought-iron gate closed behind us with a squeak, rusted by the salty ocean spray.

"No way," said Jorge. "Too much risk."

We walked home in the gathering dusk. I kept picturing Mrs. Hillman, her tiny silhouette swallowed by a huge landscape. The American Abroad. Ostracized from her native country, from her family, from everyone she had known. Alone, unkempt, wandering along the shore of the angry, crashing Pacific. End of the line.

The nursing home by the embassy was a no-go. Patients barracked in a warren of rooms, many bedridden, a slippery marble staircase leading up to the only available room. I was getting desperate. There was one last place to try. The small nursing home run by Dr. Rodríguez, the round-faced doctor whom my father had "fired" back in March.

"Call him," Jorge said. "I bet your father wouldn't even recognize him now."

Red Petals from a Mimosa

A sturdy door made of dark wood panels, each square ornamented with brass studs. A small rectangular window at eye height, shuttered. Below it, a sturdy bronze knocker shaped like a lion's head. Sounds of Pavarotti floating out of an open window above. We pressed the metal buzzer under the card with the word *timbre* (bell) printed on it and waited.

We were in San Isidro, Lima's ritziest district, two blocks from the colonial-era Parque El Olivar. Like most homes on the block, this one had a *wachimán* (guard) out front—a middle-aged man parked on a white plastic chair, an over-turned milk carton next to him, an old Labrador dozing at his feet. The dog didn't even stir when we walked up the sidewalk. The house itself was huge: two stories high, nearly a block long, painted cream with brick trim. A 1960s interpretation of Spanish colonial architecture. Low shrubs, red geraniums. Blood-orange petals dropping from mimosa trees. No sign advertised the business inside. The only clue that this was a nursing home: a wooden ramp leading from the sidewalk to the front door. For wheelchairs.

The shutters in the door-window slid open: bright-brown eyes, a round face. Jorge explained why we were here.

"Ah, yes, Señor Vera, please enter," she said in Spanish.

We passed through a vestibule and into a spacious lobby; a broad wooden staircase on the right curved to the second floor. Gathered on the sofas at the center of the lobby were five or six elderly residents, each with a white-uniformed aide by their side. Some of the residents were watching a program on a small TV; others appeared to be snoozing. A small, hunched-over man tracked our every step with his bright, mischievous eyes.

"It is almost lunchtime," the round-faced woman explained. "They are all waiting for *almuerzo*."

The bright-eyed man called out: "¡Bienvenidos a mi casa! [Welcome to my house!]."

"Gracias," Jorge said.

The man had a single tooth, dead center in his bottom jaw. How did he chew, I wondered?

The aide led us through two large rooms to the office of the director. Señora Estrella was a capable-looking woman, fiftyish, with warm, intelligent eyes and a pair of reading glasses dangling from a jeweled chain around her neck. We went with her on a quick tour: Here was the dining room, the kitchen (cramped by American standards), the other dining room, now a game room. A front living room with a fireplace was now partitioned into living quarters for three residents. Next to it was a closet-sized room—vacant, if my father wanted it. Most of the other residents lived upstairs, she explained as we mounted the stairs. There were seventeen residents total, nearly all with dementia. Dr. Rodríguez made the rounds each morning; the doctor's visits were included in the monthly rent, along with meals. Yes, it was like a family home. They preferred not to advertise. Just word of mouth. Many of the residents came from—she paused at a landing on the stairs—*well-known* families.

To our right was a paneled door; Señora Estrella gave it a push, and we entered a large wood-paneled room, the front end lit by an enormous floor-to-ceiling window. Bushy mimosa boughs swayed in front of the glass, casting shade on a four-poster bed. At the other end of the room was an ample sitting area and a table for meals. I had a pang. This room would be perfect for my father. No, it was taken. *La señora* was downstairs, having lunch. Perhaps we heard her opera music through the window earlier?

We toured the second floor. With each room Señora Estrella showed us, I became more depressed. This one was nice and airy. This one had its own private bath. But all were taken, apart from the teeny space downstairs.

We stopped at an open-air patio at the back, the floor covered with red-and-white terrazzo; tables and chairs were set up so residents could take some sun. Señora Estrella led us to the railing overlooking the garden below, which was framed with glossy green shrubs. The garden was home to a *conejo* (bunny), she explained.

I doubted it but let her go on with her spiel.

"His name is Ricardo, after Ricardo Palma."

The Peruvian poet and historian, I remembered.

We turned back and passed a door on the left.

"What is this?" I asked, trying the handle. Before she could answer, the door gave way.

Inside it was dark and musty, with heavy curtains pulled over the windows. The floor under our feet was unfinished concrete.

204 | MELTED AWAY

"It is a storeroom, señora," she said, haughtily, jiggling her neck chain.

I pull the cord to the drapes; light poured in the louvered-glass windows, illuminating a wall of mahogany closets and a jumble of furniture in the far corner. The space was large, about the size of my father's room at our house. Correction: our soon-to-be former house.

Jorge and I looked at each other.

"Would it be possible to have this room?" he said, voicing the question in my head. "We can clear out the junk, do it over."

I was envisioning carpet, new drapes, my father's recliner.

"But Señora Estrella says we have to hire theirs," said Jorge, muting his phone.

"Why?"

"Because that's how they do it."

"Jesus, Peruvians and their rules."

We were sitting in our bedroom, trying to work out the deal by phone. Señora Estrella had just informed us we could redo the storeroom and move my father in December 1. One day after Jaime's deadline. It was the best we could arrange.

"No, I don't want to use their *enfermeras técnicas*," I said, thinking of Daisy and Alma. "He has to have his own. Explain that they already know how to communicate with him, that he doesn't speak any Spanish. *Es americano.*"

A half-hour later, Jorge and I entered my father's hallway. Salsa music was pouring out of his doorway.

Daisy was facing my father, her hand on his shoulder, his hand on her hip, as she guided him in the steps. Her hips swayed as she stepped side to side in her bare feet. "One, two, three, Mr. Jhon. One, two three."

Clad in his Goldtoes, my father's feet shuffled left and right, a few beats behind the rhythm. His eyes were glued to the ground.

"¡Muy bien!" she laughed. "¡Ah, señores! I didn't know you were there."

My heart contracted. Thank god we had just negotiated the deal. Daisy and Alma had nursed my father back to health. They deserved to go with him to the *casa de reposo*. They were his family now. And they needed the money. Too bad we couldn't get a permanent job for Señora Lucinda too. She had saved him from the abyss. I didn't know how she would take the news. We would keep her on through move-in day, though.

"Daisy, we have something to tell you," Jorge began.

"Sapo verde"

It was the last Tuesday of November, toward the end of lunchtime, and Dario and I were grading papers at the back of the staff room. I heard chairs scraping, and suddenly people were flocking to the front: a table had been set up with a big white cake and bottles of neon-yellow Inca Cola.

"Shit," I said to Dario, scanning the room. The escape route to the staircase was blocked by a horde of hungry-looking copy room underlings armed with forks and paper plates.

"What's wrong?" he asked.

"I, I need to—"

"Oh, my dear. Get in the spirit. *C'est obligatoire.*"

I stood and glumly filed forward. About forty other teachers had gathered around the cake. Nelly, the cherubic fortysomething woman who ran the teacher's lounge, was sticking candles in the icing and rousing the few sourpusses still left at their tables. The lights went off.

"Who had a birthday this month?" Nelly called out in Spanish. "Please come forward. Profesor Guerra, Profesora Quijano . . ."

The chosen professors lined up by the cake, beaming like schoolchildren. Nelly lit the candles and warbled the first notes in a thin soprano.

"Feliz cumpleaños," everyone joined in with gusto. I mouthed the words. The song was followed by "Sapo verde [Green Frog]," a deliberate phonetic misinterpretation of the English "Happy Birthday." It was considered funny. By Peruvians.

Sapo verde sounds nothing like happy birthday, I mulled silently. *There isn't even a th sound in verde. Of course, they don't recognize that. Th isn't even a meaningful phoneme in South American Spanish. D. Th. Oh, who cares?*

Pieces of cake with vanilla icing got passed around, and then the dreaded queue formed: We were all supposed to line up to kiss the birthday professors, one by one.

I was dying inside. It was so juvenile. All I wanted to do was enter the final grades in my gradebook and get out. It was almost two o'clock already, time for my next class. I reached behind me to feel whether I could squeeze my not-skinny self through the chairs.

"What is your problem, Barbarita?" asked Dario. He pulled me into the line ahead of him. "They will be offended."

He was right. With everyone in the teacher's lounge looking, I would be pegged as rude if I didn't kiss—or air-kiss—the birthday boys and girls. I had learned this the hard way. My first year in Lima, at a Sunday *almuerzo*, I had backed away from kissing one of Jorge's older relatives who was wearing a neck brace, not wanting to injure her further, and she had never forgiven me. How was I supposed to know it was bad manners to be concerned about someone's spinal column?

Thankfully, there were only six birthday professors to congratulate today. I went down the line. One young woman had firm, round cheeks with pebbly acne. Another smelled of onions. The next one's skin was nearly pore-less, making me jealous. Number four had a scratchy beard. The tall camel-eyed math teacher on the end smelled delicious, for some reason. Even better than Jorge. Why did I need to know that, I wondered? In the United States, you didn't get close enough to strangers for their scent to be left lingering in your nostrils.

I rushed to the table to pack my things. Out of the corner of my eye, I watched the rest of the birthday tributes. The teachers were smiling and taking selfies with one another, as usual. Everybody was going to be late to class because of this infantile nonsense. What was next, I wondered—a singalong with Barney the Dinosaur?

Dario was next to me, gathering his papers. He glanced at me thoughtfully. "It will be your turn one day."

"It already was. In April."

"What did you do then?"

"I took my things and walked out, before they lit the candles."

I remembered the scene. People had stared. I had run up the staircase and into the tiled courtyard, the strains of "Feliz cumpleaños" wafting up the stairwell.

"Oh, Barbara." Dario shook his head. "Barbara, *que bárbara*. You have a lot to learn about a lot of things."

The Con

November 28, 11:45 a.m.

The brown leather recliner teetered sideways over the railing, threatening to plummet to the lobby below.

"Up here, to the right," Jorge called in Spanish from the second floor.

The man in front of me heaved the recliner upright as his teammates scrambled up the final stairs and took a right to my father's room. I was following several paces behind, my nerves on edge, carrying a laundry bin piled high with freshly washed sheets.

"Put it there," Jorge said to the men. "In front of the TV. To the left."

I squeezed past Jorge and stopped at the threshold of the former storeroom: The resemblance was uncanny. Gray-green walls, beige drapes, his bed positioned against the far wall next to the big window: Jorge had reproduced every detail of my father's room on Paula Ugarriza, down to the bust of Mahler on the side table.

"You made it happen fast," I said. "How did—?"

"I'll explain later," he said, tossing me a bottom sheet. "We just have to make this bed and get out of here."

The plan was coming together, I thought, fitting the sheet under the mattress. We had found the solution on a Mayo Clinic web page about how to avoid trauma when relocating an Alzheimer's patient: *Before the move, make your loved one's new room or space look and feel as familiar as possible to ease the transition.* We were taking that advice literally.

Jorge and the workmen had spent the last five days remodeling the space: painting, laying carpet, getting the cable service installed. This morning at nine, the moving van pulled up in front of our house. That was the signal for Daisy and Señora Lucinda to whisk my father into the backyard so the movers could descend on his room. In breakneck time, they packed up everything—the bed, his recliner, his CD player, artwork, clothing—and loaded it into the truck, unbeknownst to my father, who was enjoying a second breakfast behind the clothesline. Two and a half hours later, Jorge and I arrived at the Residencia San

Martín and were guiding the movers through the front door, past the gaping residents and up the twisting stairway. Now all that remained to be done was to tidy his room and return home before my father tried to enter his now-empty bedroom and realized something was up.

This plan had to work. It had to.

I smoothed the comforter and ran down the staircase, nearly colliding with the one-toothed man who had welcomed us the other day. He caught me with his outstretched arm, as if we were about to tango.

"Tell your *papi* I look forward to the honor of meeting him," he said solemnly. News had obviously spread. He released his grip and either had an eye spasm or winked at me.

Twenty minutes later, Jorge and I were in the backyard at Paula Ugarriza, behind a line of sheets that had been strung up between the patio and Pizarro's fig tree. Señora Lucinda had set up a folding table and was serving my father a sandwich and some fruit—was this lunch or his third breakfast? At his feet panted Lola, hoping for a bite. Daisy was sitting on the grass, braiding some stalks into a bracelet.

"Well, John, how do you like having lunch al fresco?" Jorge asked, ducking under the clothesline.

"It's very nice," he said, finishing a second banana.

"Ready," Jorge said in Spanish to Daisy and Señora Lucinda.

I met their eyes: We all knew what was supposed to happen next. Every element, including the phrases we used, had to go like clockwork to fool my father and keep him calm.

Daisy tossed aside the bracelet and stood up, brushing the grass off the back of her jeans. Señora Lucinda, dressed in white, stacked the dirty dishes in a plastic bin. Today was her last day; we were giving her a month's severance pay and good references, so hopefully, this wouldn't be too hard on her. When she wasn't looking, I had put some extra money and six boxes of Swiss Miss in her tote bag.

"Come on, Mister Jhon," said Daisy in English, offering him her arm.

He rose, and the two women steered him through the sheets and into the house.

"Stay," Jorge said to Lola, tossing her the uneaten sandwich crusts.

We followed my father and the women down the hallway, past the closed door to his room. Daisy disappeared with him into the bathroom.

When he emerged, my father turned left to return to his room for his early-afternoon nap.

Señora Lucinda spun him around. "*Vamos* [Let's go], Mister Jhon."

"Yes, Dad, let's go for a little walk this way," I said, propelling him into the dining room. Do-si-do your partner.

"Okay," he said. He was in fine spirits today, praise be the God of Alzheimer's Patients.

"You know, Dad," I said as we crept past the dining table, Señora Lucinda on one of his arms, Daisy on the other. "Today is the day you have a follow-up visit with the doctor. To check on your stitches." Dr. Rodríguez had removed the stitches from his head two weeks ago, here in the dining room, but he wouldn't remember that.

"Yes," I continued. "It is an important visit. We'll see Dr. Rodríguez. You remember him."

"I don't remember him," my father said, inching along. There was just so much you could do to rush an eighty-seven-year-old man.

We puttered into the living room, where the front door had been left ajar; the Volvo was purring at the curb, Jorge behind the wheel.

"Of course, you do. He has"—Señora Lucinda and Daisy escorted him smoothly down the front steps—"a curly brown beard," I finished as I closed the gate behind us.

He was now in the back seat, Daisy and Señora Lucinda on either side of him; I closed the car doors, leaped into the front seat, and we were on our way.

The women entertained him on the twenty-minute drive.

"Mister Jhon, monkey," teased Daisy.

"Daisy, monkey," my father laughed. He didn't notice we were leaving Miraflores and speeding down Avenida Comandante Espinar, toward the district of San Isidro. Past the high-priced Roe Lab that sent a nurse to draw blood at your home on a Saturday if you needed it. Past the busy Ovalo Gutierrez intersection, with its multiplex cinema and bookstores and the fancy Wong supermarket that used to be some family's mansion. A right turn at the cathedral shaped like a high-arched parabola, past the school of Our Lady of Whatever, a left at the traffic light, a quick right onto Calle Agustín de la Torre Gonzales. Suddenly, it was all hushed and orderly like a scene out of Mario Vargas Llosa's fictionalized boyhood Lima, taciturn *wachimen* stationed in front of the large two-story houses and Andean maids sweeping the sidewalks and gardeners making the suburban desert preposterously verdant.

Jorge pulled up under a mimosa tree.

"Here we are at the doctor's office!" I announced as a shower of red petals fell on the windshield. Jorge, Daisy, Señora Lucinda, and I piled out. The snoozing dog next to the *wachimán* opened one eye, peered at us, and shut it again.

Jorge set the wheelchair next to the car and lifted my father into it. Daisy wheeled my father up the wooden ramp and through the open front door, where Señora Estrella, forewarned via phone, was waiting. Seven or eight residents were gathered behind her.

"Él es americano [He is American]," a woman in a bun declared loudly.

"Bienvenidos, Señor John," said the one-toothed man. He held out his hand, which my father shook.

"Are you the doctor?" my father asked in English.

There was no elevator, so Jorge, Daisy, and I lugged my father and his wheelchair up the staircase as quickly as possible, before he realized this wasn't a doctor's office.

"Oof," said Daisy, heaving my father up the final stair and depositing him on the second-floor landing.

Señora Lucinda was close behind with a Tupperware bowl she had brought from the house. "It's special for your father," she explained, handing it to me. I lifted the lid: her heavenly *ají de gallina*.

She turned to Daisy. "Please? Can I?" she asked, pointing to the wheelchair. Yes, Daisy nodded.

A resolute look on her face, Señora Lucinda gripped the rubber handles of the wheelchair and pushed my father down the corridor to his room.

"Mira, Mister Jhon, tu cuarto [Look, Mr. John, your room]," she said, steering him inside. There was his bed, his embroidered coverlet, his reproductions of Gauguins and Manets. A Mahler symphony was playing softly on the CD player.

"Hey, John, why don't you take your shoes off and relax a bit?" Jorge said, patting the seat of the recliner. "It must feel good to be home after your trip."

"Hmmm. Okay."

He said nothing about the doctor. He said nothing about this being a new place.

He let Daisy lower him into the chair, which was facing the TV. He picked up the remote and pressed ON. A telenovela began to play. Somebody must have been in here already. I switched the channel to the Discovery Network in English. He leaned back against the cushions and sighed deeply.

A drop of mucus was dangling from his nose. Señora Lucinda took the tissue tucked inside her sweater sleeve and dabbed it. She folded the tissue into a neat square and placed it on the nightstand. She stood there for a moment, looking around.

She leaned down to lightly kiss his papery cheek. "Cuídese [take care], Mister Jhon," she said in a softer voice than usual.

Then she was gone.

"We did it," I said that night.

Jorge was lying on top of our bed, naked except for a towel, eyes closed. His chest moved up and down. No answer.

"Okay, *you* did it," I said louder. "You and the movers. I did my part pushing him up the stairs, though."

I surveyed our bedroom: Boxes were piled along the walls; my filing cabinet with the aborted reporting and book projects was sealed shut with duct tape. On Friday, our own movers would come; we had wrangled five extra days from Jaime after he was informed that my father was going into a rest home. What a disappointment Jorge's old school friend had turned out to be. For ourselves, we'd found a two-bedroom apartment by the Malecón. One-tenth the size of this house. Newer. Brand-new. Eight hundred dollars more a month. Lima was getting more expensive. Still, thanks to Jorge's efforts, none of us was homeless.

"Thanks for everything you've done for my father," I said, reaching out to stroke his graying temples. He was fast asleep. Snoring.

Illuminating his cheekbone was a crescent-shaped pool of light, cast by the streetlamp outside. I rose to peer down at Parque Leoncio Prado. Four or five teenagers were lounging on the swing set, drinking beer and cursing. On a park bench near the sandpit, a couple was making out. At this hour of the night, the park belonged to the young. An empty bottle crashed on the concrete—the very walkways where I had once imagined taking Sunday strolls with my repentant, docile father; where I had foolishly imagined us being a real family again; where he had seen ghosts and dead people and only recently had gotten well enough to take afternoon walks with his aides.

Now he had become one of the park's ghosts. A spectral figure tossing a ball to Lola, shaking hands with Will under the watchful eye of Leoncio Prado in battle uniform. My father would never walk these glossy worn sidewalks again. I knew it. Just like I knew he would never acknowledge his past cruelty, would never explain why, would never say he was sorry. All that was in the past, as Jorge frequently liked to remind me. If there had ever been any hope of an apology, Alzheimer's had wiped those cards off the table. Oh well.

I inhaled the warm springtime air, the flat, dusty Lima smell. The harsh glare of the streetlamp felt like it was my witness.

This is it, I told myself. *No more waiting. Now I just have to keep doing what needs to be done to keep him safe and healthy. For however long that is.*

I closed the curtains and climbed back in bed with my best friend.

Taken at roughly the same spot, these photos show the recession of Qolqepunku Glacier in one year: 2008 (*top*) and 2009 (*bottom*). Photos by Jorge Vera.

FIELD NOTES

The Climate Change Con

May 30, 2009, Sinakara Valley, our third Qollyur Rit'i pilgrimage

They say if you come to Qoyllur Rit'i three times, El Señor grants your deepest wishes, and I am certainly willing to open my heart to this possibility, but someone else in this tent is not.

It is our second night camping below the glacier, with two more days to go, and we're zipped in here at nine o'clock, pilgrims stomping past our campsite, drums beating nonstop, rounds of dynamite going off on the mountainside, and Jorge is having a meltdown.

"What is the point?" he says, shoving mud-soaked socks into a duffel bag. "I got all the pictures I need of the dancers; the ice is practically all gone."

"I'm going to interview that priest I contacted, Father what's-his—"

"Who cares? You aren't writing this for any newspaper."

"I can sell the story this time. I have an angle. They say the priests are telling the pilgrims the glacier recession is their fault."

"Hah!" His bloodshot eyes look maniacal. "This whole scene is too weird, and nobody outside Peru cares, and I want to go home now." He stands up and unhooks the flashlight from the roof of the tent and aims it at his open camera pack.

"No, please!" I grab his leg. "We spent all our money getting up here, and it's too soon. I have to get my interview."

"I'm getting out of here now! Tell Paco we're going."

I plead with him for an hour, shivering in the cold. Tonight it will plummet to minus ten degrees Celsius, a historic low. Undoubtedly, the pilgrims who sleep outside wrapped only in blankets will suffer. They huddle together on tarps, chewing on coca

leaves. Some will die. I can't bear the thought. The civil defense doctor's words from three years ago come back to me. "This is our way."

The next morning, Jorge follows me around the muddy site, sullen and exhausted. It's hard enough breathing in the thin air and hunting down interview subjects, let alone hauling along a resentful photographer. I feel like an evil taskmaster, but I need photographs of whatever I encounter, in case I write about it. Surely there is something here that would make a good peg for a new story—although what could be more newsworthy than a collapsing system of tropical glaciers that has implications for the entire planet, I don't know.

As I plod along, peering up at the top of the mountain, I'm troubled. The ice has receded even more drastically since our last visit. Qolqepunku Glacier has receded 60 percent in the last three years, according to experts from Peru's environmental and climate change ministry. It's on a "death watch," they say. The same is true for neighboring glaciers. It is devastating news for the people of the high Andes and the rest of Peru's thirty million inhabitants, who depend on glacial melt-off for 80 percent of their water supply. We should be in mourning here, at the top of the mountain. Instead, the pilgrimage is as exuberant as ever; if anything, there are more pilgrims than last year, despite Alfonsina Barrionuevo's predictions in 2007. I'm seeing lots of stylish dance troupes, decked out in elaborate costumes, waiting their turn to perform on the concrete platform outside the Catholic sanctuary.

And there is another development: I'm seeing more pilgrims being whipped by ukukus, especially the ukukus who belong to the Catholic Brotherhood. They're easy to spot with their blue-and-red jackets. Traditionally, they are responsible for keeping peace and order and hauling drunks up to the civil defense tent, but this year the whipping is cranked up to a level ten.

A hundred feet below the dancers' platform, a young ukuku is snapping his thin whip around the denim-clad legs of a middle-aged man in a green jacket. "It's your fault!" the ukuku says in Spanish. Shamefaced, the man drops the lumpy plastic bag at his feet.

"Why is he hitting him?" I ask a girl standing next to me.

"He stole the ice," she whispers. "It is forbidden."

The ukuku rips open the plastic bag and dumps the handful of ice into the mud. The semen of the apu, I think. Trod underfoot.

I look around for Jorge; he's photographing a line of women dancers, their skirts whirling in the air, the barren mountain and wisp of a glacier behind them. He glances at me as if to say, Enough? I point to the sanctuary.

Father Antonio is standing outside—he's the one I want to talk to—but he's surrounded by members of the Brotherhood. I approach with my voice recorder. "I'm Barbara—" I begin.

"No time now," he shouts.

Frustrated and angry, I duck in the sanctuary door. It's dark, and the floor is slippery, but at least it's quieter in here.

Standing by the altar is a dignified, black-haired man in a Brotherhood jacket. I approach and ask for his name.

Silvestre is fifty-one years old, from the Urubamba Valley.

"I have been to Qoyllur Rit'i thirty-two times," he tells me, adjusting the braided whip looped around his neck.

"Why are the ukukus whipping the pilgrims?"

"People are taking ice from the glacier and throwing garbage all over," he says. "The pilgrims are lighting fires for their cooking and making too much heat. That is why the ice is disappearing. It is the pilgrims' fault."

"Who told you this?"

He looks away. "The leaders."

He won't say more.

Outside the sanctuary, a middle-aged man is lighting a candle by a small icon of El Señor de Qoyllur Rit'i. Andres traveled for eight hours by truck from Puno and climbed up the mountain in the dark.

"The condition of the glacier is very sad now," he says, kneeling in the mud. "Es mi culpa [It is my fault]."

Weeks later, I exchange emails with Dr. Inge Bolin, a German-born anthropologist and professor from Canada who has authored two nonfiction books on the culture of the people of the remote Andes. She and I have been communicating for years about Qoyllur Rit'i, which she attended in 1999 as part of her decades-long fieldwork in the

region. I tell her about the alarming condition of the glacier and how the church is distorting the science to blame its disappearance on the pilgrims.

"It's a shame that the people who are least responsible for global warming—who are the people of the high Andes—are feeling the effects the most," writes Bolin. "Many of these people live in stone huts, without electricity, without cars. They walk everywhere. They don't eat meat from factory farms that create an awful lot of methane. They are not responsible for global climate change. And they have to cope with so much—extreme rains and droughts and loss of their potatoes because of the climate changing.

"For these people to be told they are responsible for making the glaciers disappear—it's such an irony."

6

IN THE OLIVE GARDEN
OF THE BILOCATING SAINT

The Bunny and Don Alberto

It was just after New Year's, and Jorge and I were settling into our new home on Calle Porta, half a block from the Malecón. The apartment was brand-new, with shiny hardwood floors and black granite countertops, the whole place tiny and compact. It was like living on a ship. When I stood on the narrow front balcony, I could taste the salt spray blowing in from the Pacific. Sometimes there were days on end when everything reeked of anchovies, one of Peru's main exports.

Our new neighborhood was the so-called Happy Barrio of Mario Vargas Llosa's youth, made famous in his early writings, but traces of the old way of life were disappearing fast. The Tudor-style mansions with their crumbling, bougainvillea-covered walls were being knocked down for cookie-cutter high-rises, like the one we were renting. On the streets, the construction racket started at eight in the morning and continued until six at night, six days a week—cranes, jackhammers, workmen yelling. I had to wear earplugs at my desk to get any work done in my office. A few weeks after we moved in, a bomb went off on a street catty-corner to us, knocking down an old four-story building for which the city had previously refused to allow a development permit. Two days later, a crew began hauling away the debris and mapping out new foundations. How much did that bribe cost, I wondered?

How different it was to hop in our Volvo, roll over the Villena Bridge, and head to my father's corner of San Isidro, just two miles away. The chaos of Avenida Pardo hushed to a *pianissimo* as we took a right onto Calle Torre Gonzales. There on that tranquil side street, the spacious homes of Lima's old elite still stood, with their insanely green lawns and deferential gardeners and expressionless *wachimen*. When Jorge turned off the motor and we stepped out of the car, we could hear our footsteps click on the concrete pavers that some poor maid had been told to buff to a high gloss, work she did early in the morning on her hands and knees. A polished sidewalk in the heart of the world's second-driest desert. It was beautiful and obscene. At the end of the day, that maid would take a cramped, two-hour *combi* ride to a dust-covered shantytown outside Lima, where her family likely had no running water and used pirated electricity. This was Lima in 2012. When I thought about what it really took to sustain a San Isidro—the exploitation of people, the waste of time and natural resources—it made me want to cry or laugh, I wasn't sure which. Simultaneously, I was acutely aware that San Isidro was fast becoming the only place in dirty, chaotic Lima that I could stand to be in. And who lived here? A few thousand privileged families and some elderly members of Lima high society, now gone discreetly off their rockers with dementia.

"Mister Jhon, mira. El conejo. Se llama Ricardo. [Mr. John, look. The rabbit. It's called Ricardo.]"

Alma led my father to the patio railing and pointed to the garden below. Crouched by the hedges was a small brown rabbit, nibbling a piece of lettuce. Ricardo munched methodically, his long gray ears flopped on either side of his furry, round cheeks. The bright-green leaf disappeared, inch by inch, into his mouth. Ricardo lowered his head and picked up another leaf, then another. When all the lettuce was gone, the rabbit sat up and gave his body a quick shake, like a dog flicking off the rain, and hopped into the bushes.

My father was watching intently.

"That's a cute rabbit, isn't it?" I said, entering the patio.

"Yes, a cute widdle bunny wabbit. What's up, Doc?" he replied.

Not bad for a guy with advancing dementia, I thought. Too bad my mother couldn't have stuck around to see this transformation. Except when he was drinking, my father was always dead serious. He had spent my childhood staring stone-faced at my mother and me on the couch, as we leafed through copies of *Women's Day* and *Redbook*, exploding into laughter. The 1960s and 1970s were a boon for outlandish products and so-called creative recipes.

"I don't understand what's so funny," I remember him once saying as my mother and I practically wet our pants over a recipe for Hot Dogs Banana Flambé. "Bananas are very high in potassium."

Ricardo the bunny was my father's daily mealtime companion, part of his new way of life at La Residencia San Martín, just blocks from the ancient olive grove planted almost four hundred years ago by the Peruvian saint himself. Back at our house on Paula Ugarriza, Señora Lucinda and the aides had set his schedule; here Señora Estrella ran the show. A hearty breakfast at 7:30 a.m., a 9:00 a.m. checkup by Dr. Rodríguez, a big *almuerzo* at noon and a small supper—called *cena*—at 5:00 p.m. My stubborn Yankee father was now living by the old Spanish adage: Breakfast like a king, eat lunch like a prince, and dine like a pauper.

He was even drinking like a Peruvian: glasses of apple and pineapple *agua fresca* (boiled and diluted fruit juice) with every meal. He gulped the concoctions down, no complaints, a far cry from the scotch and sodas that used to sit next to his dinner plate in Gainesville, puddles of ice water pooling beneath the glass.

The only new customs he balked at were going downstairs and getting acquainted with the other residents. I knew his grumpy self didn't give a damn about socializing, but I wanted him to do it anyway; it would please Daisy and Alma. If that goal was to be accomplished, though, it would have to be done gradually. Otherwise, he might slide backward.

"Your father is afraid of the steps," Alma told me one morning.

"Don't push it," I said. "We don't want to make him afraid. You remember back at the house?"

"Yes, yes, señora," she said, hurriedly, her green eyes flashing with worry. "We will keep doing the walks here, on the patio."

I left her and walked slowly down the staircase, trailing my hand on the wide walnut railing; at the landing by the opera lover's room, I leaned over to observe the aides below. They were all in their twenties—younger than Daisy and Alma—and dressed in white uniforms, chatting and joking, and were quick to pitch in if someone's resident needed extra help putting on a jacket or getting up from a chair. Backup. That was what had been missing in our old setup. If my father collapsed here, there would be many hands to get him back on his feet, a doctor on call to check his cuts and bruises. One of Lima's best hospitals was only a five-minute cab ride away.

I had not failed in my promise to look after him by putting him in a rest home, I reminded myself. This solution was better for everybody. So why, given all the shitty things he had done to Jorge and me, but especially me, did I feel this residual guilt?

Jorge and I joined the health club in the Marriot, overlooking the Malecón and the LarcoMar mall. The membership was overpriced for Lima. But I loved the gym's hushed ambiance, its cleanliness, the enormous plate glass windows that made me feel like I was about to sail off over the ocean.

Paragliders launched from the cliffs, soaring and dipping over the crashing waves. I pedaled as hard as I could on the stationary bicycle, trying to get my heart rate up, trying to lose the ten pounds I had gained since my father's arrival in March. The paragliders sailed dangerously close to the gleaming high-rises but for some reason never crashed. I pedaled faster and faster.

Mid-January

UPC's summer semester was about to start, and I agreed to teach in the mornings. We needed all the money we could get. Afternoons were too hot for teaching: Only the administrative offices at this newly built university were air-conditioned, not the classrooms. Teachers in the United States would have refused to step foot in a classroom under those circumstances, but the Peruvian professors didn't utter a peep: This was what one did, adapt to the seasons. Besides, hardly anyone in Peru, even well-off people, had air conditioning or central heat in their own homes in 2012. The climate was mild year-round; outside temperatures rarely reached lower than fifty-eight degrees Fahrenheit or higher than eighty-five degrees. People with money escaped to their beach homes in the summertime, to places with evocative names like Pulpos (Octopus) or El Silencio (The Silence).

Reading a newspaper article one evening, I learned Americans produced approximately eleven times more greenhouse gases than did the people in Peru. All those gas-powered cars, all those appliances, heating and cooling. Before now, I had just accepted all that as part of civilization; there was no way to wean people in developed nations from their conveniences. But living in Lima was showing me it was possible to change your habits to help the planet. It just required small trade-offs in time, comfort, convenience.

This was part of how you combated climate change: by using less electricity, by sweating a little under the armpits, by taking a siesta in the afternoons.

At the translation department's beginning-of-term meeting, I learned Dario was teaching basic English classes during the summer semester; *he must need the money too,* I thought. No summer house in Asia—the exclusive beach town two hours south of Lima—for him. And we had a new faculty member: Jack, a teacher in his early forties, from California. Suddenly, I was no longer the only American in the program.

"Glad to meet you, Barb," Jack said, extending a sunburned hand. He was tall, broad-shouldered, and ruggedly good looking; I had to look up to meet his gaze. The skin around his eyes was crumpled; there was a tinge of desperation there—*closer to fifty,* I decided.

Every American who ended up in Peru had a backstory. Everyone. Like me.

"So, Barb, what's this gig like?" he asked.

"You'll like it. You'll have to put in a lot of work, but the students are very respectful and excited to learn."

"You can show me the ropes, right? I hear you're the star teacher."

I blushed. We both knew he was laying it on thick, but I didn't mind. He seemed like a decent person, and he probably wouldn't stab me in the back.

"Sure, Jack."

—✦—

Daisy seated my father at the patio table, tied a plastic bib around his neck, and brought him a lukewarm glass of *agua de piña*. He drained it to the last drop and set it shakily on the red gingham placemat.

"¿Todo bien? [Is everything good?]," she asked, dimples showing.

"Muy bien [Very good]," my father said solemnly.

Muy interesante, I thought, observing with Jorge from the hallway.

It was a Sunday morning, and my father had been living at the *casa de reposo* for a month and a half. Jorge and I were still getting to know the other *enfermeras técnicas* who helped Alma and Daisy. There was Heidi, the determined one with stiff bangs, who made her own phonetic English flashcards to communicate with my father, and Iris, the giggler, who was constantly dropping things and getting in trouble. Other than that, I had a hard time telling the aides apart. They all had long, shiny dark hair and lilting voices and a tendency to cover their mouths when they laughed. And they were tiny: although my father had shrunk a few inches, he still towered over these women, and I sensed some were intimated by the tall gringo who didn't know any Spanish—or did he? His *muy bien* just now suggested otherwise.

While Daisy was guiding a spoonful of oatmeal to my father's mouth, Heidi dropped a basketful of laundry on the floor and plopped in an empty chair next to him. She pulled a folded index card out of her bra and leaned over, trying to catch my father's eye.

"Goood mar-ni, Meester Chon," she recited loudly. "How are jew?"

He looked up, eyebrows raised. "Very good," he said, oatmeal dribbling down his chin. Daisy dabbed it away. "How are you?"

"Excelente, Meester Chon!" said Heidi. She stuffed the card back in her bra, bent over the basket, and began sorting the laundry.

Another tech called in Spanish from the hallway. "Heidi," she said, holding up a dark-blue sock, "Did you leave this in the washing machine?"

Daisy's hand was suspended in midair as she listened. My father's mouth lunged toward the spoon.

"What does it say inside?" asked Heidi.

The woman peered inside the sock at the nametag sewn inside: "Malpartida Abreu."

"Oh, it's not mine," said Heidi. "It belongs to Martha's patient, El Chalaco."

"Okay, I'll give it to her. But *chica*, you know that El Chalaco prefers fishnet stockings!"

"Fishnet stockings? Or naughty stockings?! [¿Medias de malla? ¿O medias de malo?]."

They laughed loudly, Daisy joining in. (The joke plays on the similarity between *malla* [mesh] and *malo* [bad]. Jorge had to explain that me.)

Jorge came upstairs. He had been in Señora Estrella's office giving her a new supply of pills for my father. My father now had eight separate medicines that had to be dosed at precise intervals, and the aides kept a log with a checklist by his bedside. No more journal entries about *el paciente* and his tantrums. But overseeing my father's care here was still complicated and time-consuming. Every other day, it seemed, Jorge had to rush over with toiletries and meds. Señora Estrella hounded him when my father ran out. If a cop ever stopped him for speeding and asked Jorge to pop the trunk, he would find it stuffed with adult Pampers and butt cream.

"They're talking about a patient named El Chalaco," I whispered to Jorge. "Who is that?"

"*Chalaco* is slang for someone born in Callao. I think it's the man in the lobby. The one with one tooth."

"Don Alberto?" I asked.

"Is that his name? Yes, I think I heard the nurses calling him that."

Don Alberto, El Chalaco. Oh my god, Alberto. Alberto was missing a sock. But other than that, Don Alberto didn't miss much else that went on at La Residencia San Martín. And befriending the elusive gringo was at the top of his list.

A few days later, Jorge and I entered the house to find Don Alberto and some of the residence's women sitting in the lobby. Don Alberto had a coveted room on the first floor, with its own bathroom and a front-row view of the entry ramp, which meant he could track all the comings and goings.

He waited until Jorge went in search of Señora Estrella before he pounced.

"Tell your *papi* he must come to visit me," he announced in Spanish, flashing his one-toothed grin. "Tell him that Alberto, the *caballero* [gentleman] de Callao, requests the honor of his presence in the salon."

Don Alberto peered at the four women sitting nearby on the vinyl-covered sofas, next to their aides. "We *caballeros* have to stick together," he said conspiratorially. "Too many hens, not enough roosters here. Isn't that right, Doña Olga?" he said loudly.

This to a plump woman with a bun in a black cardigan and pearls who was snoozing on a loveseat. She opened one eye, startled. "What?"

"Oh, go back to dreaming about your lovers!" he said and began coughing. His aide patted him on the back. "Shhhh, Don Alberto. Let her be in peace."

"But you," he said to me, once his coughing fit ended, "señorita, you are beautiful. Like a dove. *Mi palomita*. Are you married?"

"Yes, Don Alberto. The man I came in with is my husband."

"No, not Don Alberto. Call me Chalaco. And remember: *¡Soy un caballero bueno sin caballo, pero soy un caballero!*"

"What was all that about?" I asked in the car. "Something about good hair and horses? It's hard to understand him. He sounds gummy with that one tooth."

"It's a pun. He's quoting Cervantes," said Jorge.

"I don't get it."

"Didn't you read that copy of *Don Quixote* I gave you for your birthday?"

"No," I said, embarrassed. It was a beautiful, thick volume with gilt edges that weighed a ton. I had tried several times to read it. It put me to sleep.

"Cervantes is playing with the word *caballero*—it means "a gentleman" and also "a mounted knight," someone who rides a horse, a *caballo*."

"Oh," I said, still hung up on the word *caballo*, which sounded too much like *cabello*, which meant "hair."

"*Soy un caballero bueno sin caballo, pero soy un caballero*. It means, 'I am a good knight, without a horse, but I am a gentleman.' Get it?"

"Yeah," I said, disappointed in myself. I could barely keep up in this language. How much was I missing on a daily basis?

I studied Jorge's profile as he eased the car into traffic. He read only nonfiction books now, but as a boy, he told me, he had devoured the Spanish classics on his parents' bookshelves. That salt-and-pepper head was filled with references to Peruvian authors and poets I barely knew—Cesar Vallejo, Alfredo Bryce Echenique, and of course, Ricardo Palma—just as he didn't catch my references to Virginia Woolf or *Middlemarch* or T. S. Eliot.

Would Lima make more sense if I read more of these writers, I wondered? Or would my head just be filled with antiquated puns only known to dreamers and nonagenarians?

That night, I dug out the Cervantes and cracked open the cover. Inside, the black satin marker lay where I had last abandoned reading years ago, in the prologue. On the page was a sonnet dedicated to Don Quixote that began:

> You, who mimicked the tearful life of woe
> that I, in isolation, scorned by love,
> led on the lofty heights of Peña Pobre,
> when all my joy did shrink to penitence.

"What is a Peña Pobre?" I called out to Jorge. "A poor *what?*"

"A poor rock."

"Why would a rock have 'lofty heights'?"

"Maybe it's a place. Somewhere in Spain. Is it capitalized?"

"Yes." I put the book down.

I knew I should look up Peña Pobre, but I didn't feel like it. I decided it was a mountain. Where people who had gotten burned by love went to get away from it all, maybe go camping, climb a glacier.

"Palomita, I bet you can't guess how old I am." This was Don Alberto's other great topic of conversation. He brought it up every time I ran into him.

"Seventy?" I said, trying to be polite.

"I am ninety-two years old!" he crowed.

The next day: "I am a hundred." The next: "one hundred two." The day after that: "eighty-seven."

"No, Don Alberto," said his aide, combing the few remaining hairs on his head. "You are ninety-seven."

"Ninety-seven? That's what I said!" He leaned forward on the couch and, with the help of his aide, stood up, shakily: "Me llaman El Chalaco [My name is El Chalaco]. ¡Soy un caballero bueno sin caballo, pero soy un caballero!"

I was grading papers with Dario in the teacher's lounge when Jack plunked a guitar case on the table.

"Oh, so you play," I said.

"Yeah, it's useful in the classroom."

"You sing songs for the students?"

"Sure, I teach them, and they sing along. It's a great way to learn English."

He snapped the case open and pulled out a battered guitar with a tie-dyed strap.

"I was in Mexico for a few years, with the Peace Corps," he said, tuning up. "I taught with music there. *If I had a hammer, I'd hammer in the morning,*" he sang. "See? It's perfect for the second conditional."

His singing voice was a lovely, rich baritone—smooth, trained perhaps. And he could really play.

"Did you study music?" I asked.

"For a couple years at Berklee. I dropped out, hit the road . . . ." He laughed in a forced way. "And—ta da!—here I am!"

"'Hit the Road, Jack,'" said Dario. "This is a song by Ray Charles, right?"

"Yes!" said Jack. "Originally in A-flat minor, but I do it in A minor." He started strumming. Dario joined in.

They're going to love him, I thought, enviously. *The students are absolutely going to love him.*

"Why did you move to Peru?" I blurted out when he finished.

Some of the light left his eyes. "My wife left me five years ago. Left me and our son, I mean."

"I'm sorry. How old was your son then?"

"Two. Yeah, he had just turned two."

"She just up and left?"

"Yeah," he said, looking at the guitar strings. "I've been raising him on my own. In Mexico. Then I met Isabella. She's from Lima. She's been great with Max."

So that was it. A former musician, emotionally fucked-up, maybe just a little bit too nice, getting up there in years, rescued by a decent Peruvian. Like me.

I met his gaze. "I get it. I married a Peruvian. A good one."

One o'clock in early February. I was standing on the fourth floor of D building, the tall one with the bright-red metal staircase outside. During earthquake drills, the staircase shook violently as hundreds of students rushed down it at once. It was almost as terrifying as an actual earthquake, which I had experienced my first week on campus. A 5.0. Everyone else took it in stride. I had felt like throwing up.

I peered over the railing; on the sidewalk below was a group of English teachers leaving for lunch. I leaned back in, hoping they hadn't seen me. A gust of hot summer wind chased through the open corridor, lifting my hair so it swirled around my face.

I was peeling a strand of hair from my mouth when my phone lit up. A Cusco number. An emotion I couldn't even name stirred.

"Paco!" I sat on the top stair, not caring if the few people left in the building had to walk around me. I had learned they would. All the Peruvians did it.

"Señora!" he shouted. "How is Señor Jorge?"

"He is well. How is Blanca, the children, the . . . alpacas?" I couldn't believe how nice it was to hear his raspy voice. I pictured his big jaw, the gaunt face always smudged with dirt, the brown felt hat with the beads and colored pom-poms.

"Todo bien," he said. There was the sound of rushing wind. Maybe they had gotten cell phone towers in Upis?

"Señora, do you make Qoyllur Rit'i this year?"

I was so still, I could feel the staircase humming underneath me. The wind chased dirt and a gum wrapper through the slats. I hadn't realized I still wanted this.

Down on the lawn, I spied a few stragglers. Most of the classrooms were now empty. *Almuerzo.* The pivoting hour.

"Señora?"

"Paco, I need to think," I said carefully. "Can I leave a message for you with Dino?"

The Date

Maybe the mountain spirit was calling me, I thought the next day as Jorge weaved the Volvo through the chaotic traffic. Another car pulled into our lane; Jorge cursed and swerved. *Maybe Apu Ausangate wants—Jesus Christ, that bus is close—wants me there.*

"Paco called me," I said out loud.

"Pucha de madre [mother of a whore]," Jorge cursed, braking at the light. "Yeah? What did he want?"

"The usual."

"What did you tell him?"

"Nothing."

I stared out the windshield. Two little kids were juggling on the crosswalk. Blue rubber balls. A rusted bucket for tips. I was seeing blue, the ice-blue heart of the glacier, and imagining the chill of that dripping down my throat, into my chest. That was when Paco had found me at the terminus and given me a hunk of his forbidden glacier ice, the first time Jorge and I went to Qoyllur Rit'i, five long years ago.

The blue chill had lodged in my heart and never left me. I just realized that.

We heard it as we trod up the wooden ramp: a high soprano warbling an aria, maybe in German. The notes trilled and echoed off the parquet flooring behind the closed door, as if the entire building were a giant music box. We pressed the buzzer; the singing continued for a few minutes, then stopped. Finally, there were footsteps, a tech peeked out of the little window; the door flung open, we entered.

Señora Estrella smiled as Jorge knocked on the open door to her office: "Oh, hello, Mr. Vera, Mrs. Vera, what a pleasure to see you." Kisses, one cheek, then the other. Lavender smell. She glanced at the parlor across the hall, where an elderly woman with a black beehive was sitting very erectly on the couch. "I would like you to meet someone."

"This is Doña María ———."

It was one of those old Peruvian family names that date to the founding of the Republic. I only knew that because the surname had appeared on one of my class rosters, and a nosy teacher from San Isidro had peeked at my list and gone into a tizzy. When I admitted I didn't know that said student's grandfather had once been president of Peru, she was scandalized. "But of course," the teacher said, "with a name like that . . ."

Now the same name had popped up in the grand old Residencia San Martín. Just in case, I would assume this well-mannered resident with black hair combed high on her head was part of the same influential family and try not to make any of my usual verbal faux pas in my new language. I hoped this conversation would not require me to say the word *trout* in Spanish, which was too close to the slang for "vagina."

Doña María had a stack of vintage LPs on her lap: Leontyne Price, Placido Domingo, Maria Callas. She showed them to us one by one, the worn album covers testament to years of listening. This one was her favorite, she said, Luciano Pavarotti. She heard him perform live twice, once in Lima, once in New York. Did I like opera? A bit, I said. I used to play the oboe. Doña María looked confused. Jorge interrupted to correct my pronunciation: *o-BOH-ay*. It felt strange to be talking in Spanish about classical music. Once I had given my life to it, but that was decades ago, before I knew a single word of Spanish. Now I was struggling to keep up my end of the conversation, lacking even the vocabulary for musical instruments.

My father? Did my father like opera? That I understood. "Yes," I said. "But he loves symphonic music. Mahler, Mozart, Beethoven. Those are his favorites."

"The Germans," she said, approvingly.

Señora Estrella chimed in: "You know, it might be very nice for Doña María and your father to get to know each other. They have much in common."

Doña María looked demurely at her lap. Glittering on her fingers were two gold rings with massive stones, an emerald and a topaz. "I heard Maria Callas in Covent Garden, in 1965," she said quietly. "The mother of Queen Elizabeth was there."

— ✦ —

As we were leaving the building, I paused by the entryway. The resident across from El Chalaco was yelling in his room. I had only seen him once—a small bald man with a stomachache face—before the aide kicked the door closed. Normally the door was shut tight, like it was today. Even so, I could make out most of what he was saying in Spanish.

"No. No. No. Get out of here, girl."

Muffled sounds of a woman pleading with him to take his pills.

His voice rose louder, deep and angry: "Fuck off!"

More pleading.

A crescendo: "Suck my ass!"

The invitation was sent verbally. Would my father like to meet with Doña María one afternoon to listen to music? The record player could be carried up to the patio.

I brought it up with him several times. He wasn't sure.

"Let me think about it."

I liked his saying he was going to mull something over. It implied that wheels were turning in there. Even if they weren't, it was nice to pretend.

I was sitting at my desk at my home office, which overlooked a hundred-year-old mulberry tree. The tree was ripe with fruit that would soon fall and turn the sidewalk blotchy purple and stick to the soles of your shoes.

My finger marked the place on my Day Runner, the third week of June.

Maybe I could. Maybe I could get Jack to cover my classes. Maybe I could pitch the British newspaper again? Surely, if I played my cards right, got an excellent evaluation this semester, the university would let me take off five days in June.

I dialed Dino's number and let it ring four, five, times. I hung up.

El Chalaco motioned to me one Saturday morning. He patted the seat beside him.

I looked at the vinyl—to make sure nobody had peed on it—and sat down. His aide was off folding laundry.

"Yes?" I asked.

He was done up nicely: a freshly pressed shirt, gray wool trousers, the few remaining hairs combed neatly over his skull. There was a dab of blood by his ear where the aide had just finished shaving him.

"Palomita, you are a beautiful woman," he began without preliminaries. "Would you like to go out with me this afternoon?"

"To the park?"

"I know a good Spanish restaurant close by. Very elegant. We'll get a bottle of wine. You like to drink wine, right?"

"Don Alberto—"

"We can go later. When the sun sets. I'll pay, of course."

Alberto's aide was standing over us, frowning. "What have you been saying to her?"

"He wants to go out and drink wine," I said.

"Don Alberto?" she said. He ignored her. "It's time for your walk. Come on."

"Just one glass. The atmosphere is very romantic," he said as the aide helped him to his feet. She handed him his hat and cane and steered him to the front door. Two residents were waiting there with their aides.

"I'm a gentleman," he called over his shoulder. "Tell your father he has nothing to worry about!"

The following Saturday, Señora Estrella gave us an update: My father and Doña María had met on the patio Wednesday afternoon. The health aides supervised the visit. They played records for twenty minutes and had vanilla wafers with *agua piña*. Neither party said a word.

Then my father asked to go to his room and took a long nap.

Dario was reveling in the spotlight. He and his lover were photographed, kissing, for the newest edition of *Hola*, Peru's glossy lifestyle magazine. A story on the growing LGBTQ movement in the country. This was a big deal. Most gay people in Peru were still closeted, like Jaime, but things were changing rapidly. Half of the country's population was now under the age of twenty-five, and many were eager to advance human rights. Dario himself had lived in the free air of Brussels for nearly eight years. Since returning to Lima, he had become an activist and a role model to many young people, including some of our translation students. They didn't envision themselves living with their mother as a so-called bachelor for the rest of their lives, like Jaime.

President Humala opposed same-sex unions; people were holding rallies in the streets, with the gatherings growing larger every week. They were being televised.

One of Jorge's nieces had just come out, and her father was having a hard time of it. He refused to let the daughter's girlfriend in the house. The next time I saw the two women together, they were on TV—at a rally in downtown Lima, waving a rainbow flag. Hundreds of people were gathered around them. The first speaker at the mic was Dario. He looked trim and handsome in his khakis and black T-shirt, a stern look on his face.

"My friends and comrades," he said in Spanish, shielding his eyes against the sunlight. "We must be free of tyranny. We have the right to love whoever we love, to be gay, to be proud of our love. Do not let the church tell you what we do is a sin. The church is an outdated institution. We will triumph!"

The crowd cheered.

My god, I realized. My work husband was a natural onstage. He oozed integrity and charisma. ¡Viva la revolución!

"I can't keep up with all the requests," Jorge said angrily. He was looking over our bank statement. "Those *enfermeras técnicas* are going through a jar of butt cream a week. Do you think they use it on the other patients too?"

We were lying on our queen bed, which filled the tiny bedroom. We had to keep the blinds permanently closed; otherwise, the people who lived across from us on the narrow street would be able to see every move we made. I hoped we didn't cast shadows on the blinds like Balinese puppets.

"Hmmmm. Maybe. It is suspicious," I said, angling my phone away from him. I was checking plane fares to Cusco in June.

"Your dad's doing okay with the rent from the tenants. His pension keeps coming in. But we still have a gap with the Gainesville property taxes." He stood up and stretched his arms over his head. "I'm going for a run."

"Out there, with all the people?" I said, nodding toward the Malecón. It was a Sunday morning, and the boardwalk was full of upwardly mobile Limeños in new tracksuits and expensive running shoes. Some women had their maids follow behind with water bottles. The rich people with the gear got labeled "los joggers," but the maids, in their flat Keds and blue-and-white uniforms, did just as much running without the recognition. It reminded me of the indigenous porters in the Andes who for decades had helped foreign climbers become the "first" to ascend a peak. What a joke. The local people had been running up and down the mountains for millennia.

I rolled up the blinds to watch Jorge make his way down the berry-splattered sidewalk, dodging in and out of runners and strolling families, his body loose and lanky. He took a right at the corner, where the world champion surfer lived. This was what he did. Run. In his old shorts and early-model New Balance shoes. He used to run on this stretch of the Malecón in the 1980s, decades before jogging was even a thing in Peru. Back then, he was the only person on the boardwalk, he told me.

I pulled the blinds down, just in case. Then I leaned against the pillows and dialed the number.

"Yes," I told Dino in English. "Tell Paco I'm on for June 18, 19, and 20."

The Procession

Late February, early March

I edited a book for a scholarly press in the States and squirreled away the money as soon as it dropped in my U.S. checking account. I kept it a secret from Jorge. It was none of his business, I reasoned. I wasn't asking him to go. This was just between me and Paco and the *apus*. I didn't have a story assignment, and I wasn't even sure why I was going. It didn't matter.

The Pleiades were going to reappear, and I was going to be there. Nobody— not my demented father nor my time-sucking job nor my obstinate husband— was going to stop me.

"Hola, Mister Jhon. ¿Quieres un té caliente—*hot*? ¿Con miel? [Hello, Mr. John. Do you want a hot tea? With honey?]"

"No, no."

"¿*No, gracias?*"

"No, gracias."

Daisy was kneeling by my father's side with a mug of warm tea, steam rising from the surface as she held the mug out to him, a paper napkin tucked underneath. His bald, mottled head shook emphatically. No, no. He didn't see me. I was on the patio upstairs, peering through the open door, not spying— *watching* unseen. It was different, I told myself.

There was a lot to take in. Daisy's patience with a grumpy old man, her insistence that he learn the proper etiquette, his acquiescing to it, but above all, something that confounded my understanding of the disease that was gnawing through his brain:

He was communicating with her in Spanish. Listening. Understanding. Speaking.

It was a mystery that defied my attempts to understand it, like the one that visited my mother at her death. She had spent the last twenty-five years of her life crippled by rheumatoid arthritis. The fingers on her hands were so deformed, they splayed from her palms at forty-five-degree angles. Monkey hands. Her joints swelled in red, bulky knobs. Toward the end, she could barely pick up a spoon. She died of stage 4 lung cancer in a Gainesville hospice on June 22, 2004.

I was with her that morning, crying and dripping water onto her parched tongue. Around 10:00 a.m., a cleaning person knocked on the door and asked if we needed new towels. "Go away," I yelled, exasperated. When I looked down at my mother, she had stopped breathing. I laid my head on her chest and clung to her ebbing warmth, like a baby animal. About a minute later, I raised my head and glanced at her fingers: They were long and straight and without the disfiguring red knobs. It was like the rheumatoid arthritis had never happened.

"It's the Grace of God," said the nurse on duty, who had snuck in while I was bawling my eyes out.

I pointed to the fingers: "Have you ever seen this before?"

"No," he admitted. "He does work in wondrous ways."

Was it the grace of god, the double-capital Grace of God, or just some kind of freaky phenomenon that science couldn't explain?

Two weeks after I not-spied on Daisy and my father, Dr. Aguirre sat me down with a pencil and a diagram of the human brain to explain what was happening.

"Alzheimer's is a mysterious disease. It progresses differently with everyone. It affects this part and this part and this part . . . ." He drew circles around the frontal lobes, the hippocampus.

"But *here*"—he pointed with the eraser tip to the back of the brain—"here is where we process incoming language. Spoken language, written language. And Alzheimer's often does not affect this part of the brain, Wernicke's area, until the later stages. So, your *papi*, even though he is forgetting many things, he can learn new words and even the basics of a new language. Like Spanish."

The only thing I could say was "wow."

I kept going to the gym, building up my endurance, increasing my lung power. I envisioned myself easily climbing the mountain to Qoyllur Rit'i, no more huffing and puffing. I'd be like those rugged village ladies, the *mamachas*, in their multilayered skirts, scrambling up the mountain in their rubber sandals, knocking dawdlers aside to quickly reach the sanctuary, pay their respects to the Lord of the Snow Star, and get back down the mountain by nightfall.

I lost three pounds and went out for ceviche to celebrate. Two days later, I weighed myself: I weighed one pound more than when I started. Was it even possible to lose weight when you were fifty, I wondered?

Don Alberto stopped asking me to go out with him. Now I just sat next to him on the sofa and listened to his jokes and pretended to understand them. The other residents who hung out in the lobby didn't laugh at them anymore, and I could tell he needed an audience.

Two women were always there when I visited: a delicate woman in her late seventies with a walker and the plump one with the bun who wore pearls and cardigans and was often asleep. When she awakened, her eyes popped wide open—a vivid agate blue, with indigo rims around the iris. She had the kindest grandmotherly face and a benevolent smile. What was her story? Why had her family chosen to put her here, I wondered?

"Question 3, *B*; question 4, *C*; question 5, *A*; 6, *has visited*; 7, *gone*."

Page flip. Test slid to the bottom of the stack. Next test at the top.

Jack and I were working as an assembly-line team, one of us reading out loud, the other marking the test sheets. Jack had suggested the method; now that I had tried it, I was all for it. It cut grading time in half and reduced errors. We had done something similar when I was a proofreader at *People* magazine, eons ago.

Half an hour went by; we got through two sections' worth of grading, his and mine. Dario looked on as we tabulated the grades, shoved the graded tests into manilla envelopes, and wrote the averages on the front. He had no other French teachers to share grading with yet, but I suspected that would soon change. More translation students were signing up for French as their third language, and our director would have to hire more professors. As Dario predicted, Quechua was temporarily being phased out as an elective.

"Well, what did I tell you?" Jack said, once we finished. "Easy peasy, mac and cheesy."

"Yep, easy peasy," I said. Jack taught these corny phrases to our students, and they ate them up, so I found myself using them. "Cooking with gas." "Nifty." "Made in the shade." *Los idiomas americanos,* our students called them, and Jack hadn't gotten around to telling them that Americans said these things in the 1950s, not in 2012.

"What do you say we grab a bite to eat?" Jack asked me and Dario.

"Well . . ."—I had brought a sandwich. I also had a favor to ask—"sure."

The sun hit our eyes as we emerged from the underground tunnel by the university's front gates. Our lunch group had swelled to five. I skipped ahead of the others to walk with Jack. He strode along, all six feet of him, parting the sea of students streaming toward us.

"Hey, I wanted to ask you." I did a little hop skip to keep up with him. "Would you be willing to fill in for me the week of June 18? My two intermediate classes?"

"Take over your classes? I'll have to check my schedule."

He had the time. I knew it. I had arranged my class schedule for the new semester so I would be teaching during the class blocks when he was free.

At the Chinese restaurant, Jack pulled out his phone at the table and checked his calendar: "June 18, 19, 20, 21: Yeah, those dates are okay. Happy to oblige."

Dario gave me a penetrating look under his long eyelashes. "What are you doing on those days, Barbarita?" He took the pot of hot tea and began passing small white cups around. The other teachers ordered pisco sours.

"Jack is just filling in for me, that's all," I said. "I have a previous engagement."

"A 'previous engagement.' That sounds official. What type of previous engagement? A conference? A wedding?"

"Something work related."

"Hmmm . . . ." He handed a cup of tea to Jack. "You should watch out for her. She is not always level."

"On the level," corrected Jack. "She is not always on the level. But I know she is. Right, Barbie Doll? Chin-chin," he said, knocking cups.

"Chin-chin," said Dario.

"Thank you, Jack," I said. I kicked Dario under the table.

"I keep you honest," he said.

That Saturday, as I was coming down the stairs at La Residencia San Martín, I saw something that stopped me in my tracks. The woman with the pearls was sitting upright, her fingers pressing rhythmically on the crotch of her pants. She smiled dreamily as her fingers worked around and around. The TV, which got lousy reception, was blaring something about the capture of Florindo Flores, the Peruvian terrorist leader and drug lord.

There was nobody else in the lobby. I decided to leave.

A few evenings later, out of the blue, Alberto said to me in Spanish: "Palomita, your *papi* doesn't want to be my friend."

"Oh, no, Don Alberto. That's not true."

"He doesn't want to be friends with El Chalaco."

"You're wrong. He doesn't speak Spanish. That is the problem."

"He doesn't speak Spanish? What language does he speak?"

"English."

"English? He is from Germany. He must speak German."

"No, Don Alberto. He is from Florida, in the United States."

"¿Un americano?" Alberto looked scandalized. He sighed. "I am all alone here with the girls. He should come down the stairs and be my friend and speak Spanish."

I was silent.

"Tell him he should come down the stairs," he repeated.

Why was my father still not coming down the stairs?

He was walking around the second floor as his daily exercise.

He was getting over a little fall he suffered at the hands of Iris two weeks earlier. She had been helping him get out of bed and couldn't support his weight. Dr. Rodríguez had to put four stitches in his head.

Daisy was still friendly to Iris, but Alma was pissed. She refused to eat in the staff kitchen when Iris was in there. She took her soup up to the patio and slurped it at my father's table, a stormy look on her face. She gave Iris the silent treatment when they traded shifts.

"You are angry at Iris," I said to her one day.

"Your *papi* is my responsibility," she said gruffly. "I must make sure the young ones take care of him properly."

Señora Estrella called us in mid-March. My father had finally walked down the stairs! He did it backward, holding onto the railing.

"Drumroll, please," said Jorge after he hung up. We were sitting in our living room; Lola was outside on the balcony barking at passersby.

"He's old and has dementia, and this is a big deal," I said.

"Some big deal."

"Boy, are you sarcastic lately."

"Really? Well, I can't believe this is what my life has turned into, people calling me up because my father-in-law walked down a flight of stairs."

There was usually one reason he got bitter, I thought. He would never admit it, though.

"Señora Estrella told me one more thing," I said. "My father is going to join the others for their morning walk in the park."

"This I've got to see—not!" he laughed.

"By the way, did you lose out on the ARTE gig?"

He looked at me, surprised. "Yeah, they canceled the whole thing. They don't want us going into the mining camp where journalists got killed last month. Too much liability."

"Wow, I'm sorry," I said, taken aback. I hadn't realized he was gearing up for a shoot about illegal mining in the Amazon. Thank god ARTE had canceled. But this would mean less money for us.

Don't ask me about June, don't ask me about June, I thought. *I don't have a story cooked up yet.*

I had settled into afternoon teaching. It was winter again. A year ago, my father had been running around naked and trying to kill himself. Now he was part of Dr. Rodríguez's well-oiled machine, or at least that was the picture Señora Estrella painted for us when we visited one Saturday.

"Your *papi* is so . . ."—her eyes, rimmed in dark eyeliner, scanned the office walls as she searched for the word—"he is so cooperative and polite."

"Cooperative?" said Jorge incredulously. He set the leaking tub of diaper cream on Señora Estrella's teak desk and hurriedly looked around the room. "May I?" he asked, hand hovering over a box of tissues.

"Of course," Señora Estrella said. "Yes, your *papi* says good morning to the other residents when they take the morning promenade. He lets Doña María go out the door first. And he and Don Alberto are great friends."

"Great friends?" repeated Jorge, wiping the blobs of cream from his fingertips.

"You should see them together."

"I would like to see that," I interrupted. "What time?"

"Ten in the morning. More or less."

"So, overall, my father-in-law is fitting in here," said Jorge, more like a question.

"But, of course, Señor Vera," she said. "We never expected anything else from our American gentleman."

It was a Thursday morning, a little after 10:00 a.m., and I was sitting on a park bench in El Bosque Olivar (the Olive Grove Forest), waiting for signs of *el caballero americano* and his new posse.

It was a much bigger park than Leoncio Prado, at least twenty blocks long, and older, far older. It dated to 1638, when a lay brother in the Dominican church named Martín de Porres Velázquez planted the city's first olive grove with saplings from Spain to provide cooking oil for the people of Lima. That effort was so like him—attentive, humble, miraculous San Martín, the patron saint of sailors and barbers, the man who fed 160 homeless people a day with the alms he begged on the streets of Lima, the miracle worker who healed people of the plague (sometimes with only a glass of water), a mystic said to levitate in his monastery cell at night, one of the few saints reportedly able to "bilocate" (be in two places simultaneously). The nearly seventeen hundred dust-limned olive trees now dotting the length of the park had grown from saplings tended by San Martín himself and, later, by a nobleman called the Count of San Isidro. Not far from this bench used to be an old olive press; when Jorge was a boy, in the 1960s, his father drove him here to buy olive oil for the family. Now the press was gone, and it was a quiet, elegiac park dotted with squat, gnarled trees and paved with reddish sidewalks where Limeños came to walk and skate or sit on benches to dream, read the newspaper, take a nap.

I checked my watch again: 10:15 a.m. I was tempted to catch a cab to Monterrico and be done with it.

Then, coming down the path on my left, I saw them: a line of residents from the *casa de reposo*, each led by an aide in white, the bossy head nurse doing reconnaissance in front of them. They were walking slowly, some peering at the ground, others at the twisted olive trees, coming closer, closer. Leading the pack was an energetic man named Dudi, who had been diagnosed with early-onset Alzheimer's several years ago, in his midfifties. Behind him, Mr. Suck My Ass. The masturbating lady with the pearls. Two residents in wheelchairs. Doña María with her beehive, in bright-pink pants. Bringing up the rear: Daisy, with Don Alberto on one arm and my father on the other.

"Señora!" Daisy called when she saw me.

My father was concentrating so deeply on navigating the red pavers, he didn't notice when I joined them. I slipped my arm through his.

"Oh, hi, Barbara," he said, hoarsely, looking up.

"Hola, Palomita," Don Alberto said. "I have been telling your *papi* about our evening plans."

"Don Alberto, forget that idea," Daisy scolded. "Señora Barbara is very busy teaching."

"I will help you grade their papers," Don Alberto offered. "Then we can go to the beach."

"Our family has a membership in Club Waikiki," Doña María said haughtily. "That is the best beach in Lima, you know."

We inched along the promenade, past several couples on park benches, an artist sketching the trees on a notepad, a depressed-looking young woman writing in a journal. *Cuculís* droned in the dusty branches. What a motley crew we were, I thought. I could only imagine the scene we presented to onlookers. The townspeople traipsing across the hillside after Death and his scythe, in the last scene of *The Seventh Seal*. The *Totentanz*.

"Stop here!" the head nurse shouted in Spanish. She tossed her sweater on a park bench and began doing side bends. Dudi followed her every move: he was always raring to go. Suck My Ass wandered off to chase a pigeon until his aide retrieved him.

"Okay, Mister Jhon?" Daisy asked. "Exercises."

"Okay," he said uncertainly.

Arms up, arms down. Flap, flap, flap. Around and around. My father watched and joined in, stiffly. Don Alberto huffed and puffed alongside him.

"Come on, my friend," he said in Spanish to my father. "Let us work up an appetite before *almuerzo*."

Ten minutes later, they sat down for a breather. Some *cuculís* were kicking up a fuss in the branches; perhaps it was mating season again?

Don Alberto peered around my father to address me: "Palomita, do you still like to dance?"

Still? I wondered. "Yes, sometimes, I do," I told him.

He stood up and shuffled back and forth. "Mambo, cha-cha-cha."

The aides and my father laughed.

"Well, what are you waiting for?" he said.

"Not today, Don Alberto."

He winked at me. "We will dance together one day. I feel it in my, in my—"

"Don't say it, Don Alberto," Daisy interrupted. "We all know where you feel it."

Jorge came home that afternoon, excited and covered with dirt. He was renovating the bigger room in his art studio downtown, in a dilapidated colonial mansion. And he had a name for his new school: Lima Foto Factory.

"I like it," I said, as he filled the bathroom sink. "It's Warholesque."

He dunked his face in the soapy water and scrubbed with a washcloth. I met his gaze in the mirror; he was now in forward motion, no longer stuck.

"We'll have basic classes," he said, drying his face, "and one on environmental photography and, of course, nude workshops."

"Of course?"

He hung up the towel. "Limeños will eat it up. You forget. This country has missed out on thirty years of art and culture. People are dying for something different."

"Yeah, I bet you're right," I said, but I wasn't thinking about him making up for the Shining Path years. I was thinking what all artists' wives think when their husbands announce they're embarking on a nude series.

Of course. Nude workshops. I had to trust.

My own plans moved forward. I had the five days charted out. I had my substitute lined up. I had worked out a fair price with Paco. Maybe I could get Dario's Quechua-speaking translator, Nati, to come along.

I even knew why I was going this time. I wanted to touch the blue ice. Before it vanished.

It was 1:30 p.m., and I was correcting essays in the teacher's lounge. It was the last Tuesday in April, and I knew what Nelly was about to say. For whatever reason, I wasn't pissed off this time. "Would everybody please come to the front? Profesor Siegal, Profesora Drake . . ."

The lights went out. I wasn't loving this, but I didn't bolt.

"Hah, hah! Barbarita," Dario laughed. "You are the birthday girl." He and Jack pulled me up from my chair and gave me a push.

I joined the other professors around a large white sheet cake decorated with garish pink flowers.

I could feel the goodwill encircling us as people sang "Sapo verde a ti," Green frog to you. I endured their kisses; I sank my fork into the spongy yellow cake with the sugary white icing that filled my mouth with slippery grit. I wasn't running. I wasn't running. Why would I run from something that for five minutes made all these decent, unneurotic people so genuinely happy?

Calendars

Mid-May

"You are what?" He put down the razor and stared.

"I'm going to Qoyllur Rit'i in June. With Paco." For some reason, my voice had shrunk. "You don't have to come. It's all settled."

"Where is this money coming from?"

"I saved it. Plus, I have another thousand from my next paycheck."

"You do know we still haven't paid the aides from last month."

"I . . . I didn't know that." I felt an urge to run into my office to check that the eight hundred was still there in the inlaid wooden box where I stashed things. Shit. We owed money to the *casa de reposo*. I had no idea.

His mouth settled into a thin line as he picked up the razor and leaned toward the mirror. He tilted his chin and gave his jaw a good hard scrape.

Two days. Three days. He had stopped talking to me.

"No, I'm not mad," he said when I asked him. I wanted to kick his hairy shins.

I felt it radiating from him. Ice. Blocking all warmth.

Fuck it, I thought. I would just go the budget route with Paco. No burros. No extra rice and jam. No translator. Just a tent and our other stuff in backpacks, me and the alpaca farmer climbing the mountain to Qoyllur Rit'i along with all the other desperate pilgrims.

"So, it's your father's birthday next month," Señora Estrella said, looking chipper. Jorge and I were in her office, paying the prior month's bill; through the sliding glass doors, I could see the bunny munching under the shrubs. Now that it was winter, Ricardo Palma spent a lot of time in his hutch or hiding behind the clothes dryers, where it was warm.

"Yes, the fifth," I said, watching her counting out the twenty-dollar U.S. bills, the ones I had just cashed from my paycheck. Down the street were licensed money changers, wearing green vests, who changed Peruvian *soles* to American dollars, which Dr. Rodríguez preferred.

"I'm sure you have started planning already," she said, writing up the receipt.

"Of course," I said. *What's to plan?* I thought. *A little cake from Wong, a card . . .*

"What days are you going to Qoyllur Rit'i?" he asked in the car.

"June 17, 18, 19, 20, and 21. Jack is covering me at work. I have it all planned out."

"Uh-huh," he said. The iciness was gone. He usually got over these fits in a few days. I hated them.

"What did she mean back there, '*Be sure to save me a piece of cake*'?" I asked.

"I wouldn't worry," he said, starting up the car. "You know Peruvians and birthday parties."

"Señora, we are so excited!" said Daisy. She was tidying up my father's room, dusting the shelves with something that looked like two battered turkey feathers tied to a wooden dowel. My father was watching a Peruvian soap opera called *Mi amor, el wachimán*. It was about a poor boy from the provinces who becomes a *wachimán* in Lima and falls in love with the blond daughter of a rich businessman. The older man who gets him the job turns out to be a criminal who keeps trying to kill him and the girl. Sometimes they have chase scenes on rickety three-wheeled vehicles in the desert. That was how people got around in the shantytowns.

"Yes, you're excited?" I said to her, bending down to kiss my father's cheek. It had become normal. I was kissing everybody all the time here in Lima, so why not him too?

"The party!" Daisy said. She flicked the turkey feather duster over the TV screen. The *wachimán* was kissing the blondie in the laundry room.

"Well, it is just a birthday celebration," I said.

"Yes, and everybody will be there: All of the residents, the *enfermeras técnicas*, Señora Estrella. We are looking forward to it with *mucho, mucho* anticipation."

I felt the color drain from my face.

"What kind of cake will you serve?" she yelled down the staircase as I ran to Señora Estrella's office. "Vanilla or chocolate?"

"One for about forty or fifty people will be good," she said, smiling at me as she removed her eyeglasses.

"Forty or fifty people," I repeated, numbly, trying to imagine the size of this cake.

"These people will take care of everything," she said, handing me the card of a ritzy bakery in San Isidro, the one with the elaborate tiered wedding cakes in the window.

Oh well, I thought. *In for a penny, in for a pound.*

Riding home from UPC two days later, I stared out the cab window at the dirty streets and dodgy new construction and imagined myself on the trek to Qoyllur Rit'i. The rough-and-tumble town of Mahuayani at the base of the mountain where open-top trucks dropped off pilgrims from distant provinces, the noisy vendors hawking bottled water and alpaca jerky; the dusty red path up the mountainside, a thousand vertical feet up, crammed with groups of villagers and llamas and stubby-legged dogs and fast-footed *mamachas*; the wheezing in my throat as I gasped for air; after four of five hours of climbing, reaching the Sinakara Valley, about sixteen thousand feet above sea level; the straw-colored ichu grass, the lichen-covered rocks, the trickle of glacier water winding parallel to the rocky path; the stone crosses erected every several hundred feet; the pilgrims kneeling, praying, singing.

The sound of the drums. *Boom, boom, boom-ta-dum.* Over and over and over.

They said if you went to Qoyllur Rit'i three times, El Señor would grant you your heart's desire. This obviously had not happened for me, but I wasn't holding it against Him. Maybe I hadn't formulated my request properly. Maybe you had to get the wording and the timing just right, like an article pitch. Did I even know what my heart's desire was?

In the front seat, the cabdriver cursed and swerved around a strange vehicle to our right; the driver had welded a wooden wagon bed to the front of a motorcycle and was transporting his family in it.

The red-faced woman and her three children stared at me, frowning, as my driver gunned it.

I was killing time in the UPC library, waiting for sixth period to end so I could head out to D building, when I picked up a copy of the tabloid *Peru 21*.

On page 3, there was a photo of a man in an enormous sheep's head mask, dancing under an icon of El Señor de Qoyllur Rit'i.

The headline blared, "100,000 pilgrims expected for the pilgrimage."

My heart stopped.

"The indigenous ritual, which began in Inca times, will be held June 2–5."

"Sí, sí, señora," he said, over and over, when I finally got in touch.

"Why were the dates erroneous?" I yelled in stilted Spanish. I meant, *Why didn't you know the right dates in the first place and correct me when I first booked the trip?* It wasn't fair of me to blame him, I knew, but I couldn't help myself.

"Sí, sí, Qoyllur Rit'i is the second of June," he said.

"But why not tell me in January? The date?"

"The date? Oh, señora, we only know that when we are near to Qoyllur Rit'i. The stars . . ."

I could hear the wind on the other end of the line. He and Dino were starting off on a trek to Machu Picchu. That grueling climb up the Inca Trail, Paco with a seventy-pound pack on his wiry back, the tourists lumbering behind with their walking sticks, gasping for air.

Dino grabbed the phone back. "Did you work it out?" he yelled in English.
"I don't know. I don't know if I can make it."
The call cut off.

— ✦ —

It was my old weakness, a tendency to overlook details. Leave it to me, organizing my first solo trip to Qoyllur Rit'i, to get the dates wrong.

It was a lunar festival. The dates changed every year. *I should have checked, I should have checked,* I thought.

The anger boiled up like a poison. *I hate myself, I hate myself.*

— ✦ —

I was back in the teacher's lounge, savagely grading essays. Three spelling errors, and their grade dropped a whole letter grade. More than five: *F*.

Jack dropped his guitar case on the table and waved a hand over my face. "Hello? Anybody there?"

"Oh, sorry," I said, looking up at his worried face. "How are you doing?"

"Fine. It's you who seems out of sorts."

"Yeah, I've got a couple of things going on. Oh, that reminds me. It turns out I don't need you to sub for me June 18. Could you do it earlier—June 4, 5, and 6?"

"Let me see." He opened the little vinyl datebook UPC gave us. "Not the sixth after 3:00 p.m., but otherwise, sure."

"Great."

"Hey, are you okay? I mean, are you really okay?"

"Honestly, I don't know."

— ✦ —

I paid the cabdriver and got out by myself, in front of the *casa de reposo*. The solution had come to me last night as I was brushing my teeth. I could switch his party to another date in June. What difference would it make? He didn't know one day from the next. I would explain it all to Señora Estrella.

"¡Buenos días, Palomita!" Don Alberto said loudly as I stepped into the lobby. "We are all looking forward to your father's birthday party! You will dance with El Chalaco then, right?"

My father was sitting behind him, on the sofa, Doña María at his side. An emerald flashed. Was she holding his hand?

"Oh, hi, Barbie," my father said, hoarsely. "I'm going to have a birthday party with my friends. June fifth. *El cinco de junio.*"

"Your *papi* has a cold," Alma whispered to me. "We are giving him avocado honey syrup. The doctor is going to get X-rays performed on his chest."

"Will he be okay for the party?"

"Of course, he will be fine. He is so excited for his birthday this time."

I turned around without visiting Señora Estrella and caught a cab home.

There was a glacier that was melting. There was a father who was turning eighty-eight years old. There was a celebration that might be the last one he would ever have or remember. There was Paco and his family in a stone hut heated by dried alpaca poop and his mother yelling at him about strawberry jam. There was an article I desperately wanted to write and publish.

Unlike San Martín de Porres, I could not be in two places at once. Nor was I good at juggling dual agendas. I could only handle one thing at a time, I had learned.

Okay, then. I would have to choose one thing and do it very, very well.

Bésame mucho

"¡Feliz cumpleaños!"

"How many years?" asked the lady with the pearls.

"Eighty-eight," Don Alberto said. "He is still a young rooster. I, on the other hand, am 102!"

"No, you are not," Heidi corrected him. "You are 97."

"That is what I said," he said, peeved. "Why don't you ever listen to me?"

The extra dining room at the front of the house had been cleared of the surplus wheelchairs that were normally stored there; a white linen tablecloth was laid over a large oval table, set with fine china and real silverware, buffed to a high gleam. There were red and white balloons tied to the backs of chairs, fresh yellow carnations on the sideboard. A portable CD player was playing Frank Sinatra's "New York, New York," Jorge's choice. Propped on chairs around the perimeter of the room were nineteen or so residents, dressed in their Sunday finest, hair combed and neat, their eyes fixated on the large, elaborate cake, its smooth white surface adorned with blue sugar rosettes and eight burning candles.

My father sat in front of the cake, Alma standing behind him, his eyes wide and watery. His cold wasn't better, and we had asked him again this morning if he wanted to postpone the party, but he had insisted, no. Today was the day.

Don Alberto began thumping on the table with his fist. We launched into "Feliz cumpleaños," followed by "Sapo verde." I shot a warning glance at Alma, who pulled my father back from the flames. Daisy was here, too, even though she had the day off.

"Make a wish, Dad," I urged when "Sapo verde" was over.

"Go on, blow!" Don Alberto said in Spanish.

My father closed his eyes and gave a feeble puff, the flames barely flickering. I hurriedly began blowing. When his eyes opened again, the flames were all out.

I bent down and kissed him on the cheek. He looked up and smiled. He still knew me, I thought, marveling. Apart from that, he was on another planet these days.

"Well done, Señor Jhon!" Señora Estrella called.

She shooed the aides forward to start slicing and distributing the cake.

Forget the American custom of giving the first slice to the birthday boy; it was every aide for her patient and herself.

I grabbed a plate with a big slice and deposited it in front of my father. He finished it off in four bites and asked for more.

"Mister Jhon, slow down!" Daisy laughed. She pulled up a chair and poured him another cup of soda.

Ten minutes later, he had devoured three pieces of cake, two bowls of ice cream, and two Cokes. He began reaching for another slice.

"That's enough," Alma warned in Spanish. She caught my eye. "Thank you for the party, señora. You are a good daughter."

"Oh, I don't know . . ."

"No, you have done the right thing for your *papi*. You have *mucho cariño*."

"Thank you." I had heard this word, *cariño*, before. I needed to look up its meaning.

"So, this is quite the shindig," Jorge said, giving me a hug. He had an art opening at Lima Foto Factory in two days, but he had taken a break from hanging the exhibit so he could be here.

"Did I have a choice?"

"Nope. Señora Estrella made that clear. So . . . you're okay with missing Qoyllur Rit'i?"

I stopped eating my cake and tried to gauge how I felt. It was odd. I wasn't resentful at all. This was what I was supposed to be doing: celebrating the eighty-eighth birthday of my father in Pizarro's City of Kings, with a bunch of good-hearted Peruvians, several of whom had saved him from a horrible death last year. One thing at a time.

"Yeah, I'm good with it," I said. "What does *cariño* mean?

"Love, affection."

"So, like *amor?*"

"Well, yes and no. It's the sweet feeling you have when you care for somebody and do things for them."

"So, it's not just a feeling, but it's what you do with the feeling, right?"

"Hmmm. I never really thought about it before, but yeah, I guess you're right."

I helped the aides clear the table while Jorge took off the Sinatra and put on another CD. An acoustic guitar riff reverberated, and then a deep voice sang out, "Bésame mucho [Kiss me with passion]." I could practically feel the women in the room, young and old, melting inside.

Over in the corner, Doña María discreetly edged to the front of her chair and looked sidelong at my father, who was licking phantom icing from his fork.

"Hey, Dad, would you like to dance with Doña María?" I asked.

"Don who?" he said. "Oh, no. I don't know how to dance. Your mother knows that."

"My father is sorry, but he can't dance with you," I yelled at her in Spanish over the strumming guitars. "He has a cold."

Doña María turned pink. "I prefer opera," she said haughtily.

"¿Palomita?" Don Alberto was standing at my elbow. The top of his head barely reached my shoulder. "Would you like to dance with El Chalaco?"

I looked down at his impish smile. Sí, I nodded.

His hand was hot and dry in mine as he steered us in a sort of cumbia-foxtrot—back and forth, back and forth—while Lorenzo Bon sang about kissing his love as if this were the very last time. Afternoon sun poured through the intricate grillwork on the front windows, making lacy patterns on the tablecloth. I was conscious of Don Alberto's powdery smell, the steely frailness of his arms, as he marched us toward the doorway. I stumbled on the threshold as he pivoted into the hallway. It was darker in here. One of the sconces wasn't working.

He let go of my hand and caught his breath. A little spider was spinning a web in a corner of the doorframe. Don Alberto stood near, not touching me, gazing at the floor.

"Bella, you were so beautiful that night," he said in a tight voice.

"What?"

"That night in El Silencio." He closed his eyes. "With your undulant hair. Your blue crepe dress. I felt like I was falling off a cliff, called by your love."

I felt very quiet inside. All these months, his overtures.

"Remind me, Don Alberto, when was this?" I asked gently. "My memory is not so good these days."

"The second of January 1935. Your papi threw a party at your family's beach house."

I inched closer to him. "Yes, I think I remember now."

"I went into the study of your papi to ask for his permission, one gentleman to another. He said no. He said I was a chalaco with a big mouth and no money.

I did not know what to do. I ran away." His shoulder trembled through the thin shirt. "I left you completely alone."

The spider threw out another line.

"Do you forgive me, my little dove?" he asked, looking up at me.

"Of course, I forgive you, Don Alberto."

"Thank you, my love. Thank you a thousand times." The words came from deep inside.

"May I?" he asked, taking my hand.

He lifted it to his dry lips and kissed it. I wanted to cry.

"Chalaco, where are you?" someone yelled from the other room.

"Teaching this young lady how to dance!" he yelled back.

He led us to the doorway.

"Thank you for the dance, Don Alberto," I said.

"It's not over," he whispered back.

Taking a deep breath, he steered us to the center of the dining room, as Bon sang of perhaps being far from his love tomorrow, the implication for tonight being . . . . Don Alberto tilted me backward into a small dip, making my head spin. I could feel his body strain as he pulled me back up. We were both sweating now.

"¡Magnífico!" Daisy called as people clapped. My father just stared. Jorge gave me a thumbs up.

"¡Soy un caballero bueno sin caballo, pero soy un caballero!" Don Alberto said loudly.

"You sure get around," Jorge said when I made a break for the table. I started checking the bottles on the table. There was hardly any soda left.

"I think I need to sit down," I said. "Can you get me some water?"

From the corner of my eye, I saw Alma slip out of the room. The front door creaked open; through the grillwork over the windows, I could see her walking carefully down the wooden ramp, a slice of cake in one hand.

On the front lawn, the wachimán was dozing in his chair, head back, mouth open. Alma tiptoed over and set the plate on his milk carton table and covered it with a napkin. He didn't wake. Nor did the old Labrador snoring at his feet.

The front door closed again.

A light breeze stirred the mimosa trees outside. Blood-orange petals fluttered to the ground.

Alejandro Espinoza Pando (*right*), leader of a *comparsa* from Santiago de Cusco, at their campsite at Qoyllur Rit'i, June 2009. Seated *center* is the group's *ukuku*. On the *left*, below the banner, is the *ukuku*'s miniature doll self, through which he speaks in falsetto while in character. Photo by Jorge Vera.

FIELD NOTES

The *Comparsa* Leader

June 1, 2009, 8:00 a.m., Sinakara Valley, our third Qoyllur Rit'i pilgrimage

"Don't step in the donkey shit," Jorge says as my boot lands in the mess. I scrape it off on sharp-edged boulder covered in dried-up moss.

It is impossible to tell the mud from the shit on the last morning of the pilgrimage. Over the last three and a half days, hundreds of thousands of people have stomped through this glacier basin, and the campsites are a jumble of tarps and cooking fires and pilgrims milling around in their ritual costumes. The mountains looming over us have lost nearly all their ice caps, but Alfonsina Barrionuevo was wrong. The people keep coming. I can't figure out why.

Jorge, Paco, and I cross a narrow glacier stream and head west, away from the incessant drumming. Some village pilgrimage groups, known as "comparsas," are gathered on this side of the valley, their sites marked by embroidered flags that proclaim their district and the name of their "majordomos," pilgrimage sponsors and leaders. I'm drawn to an impromptu shrine of candles, offerings, and a baby doll dressed as an ukuku—considered the ukuku's animus, or vital force—arranged beneath two large banners, one made of orange silk with yellow embroidery, the other of black velvet, decorated with ornate white flowers.

NACIÓN QUISPICANCHIS, the orange banner reads. This group is part of Quispicanchis, one of the ten nations, or brotherhoods, of the surrounding area that have attended Qoyllur Rit'i and participated in its rituals since pre-Inca times. Another flag reads, SANTIAGO DE CUSCO. The poorest district in the Cusco province.

About twenty comparsa members are standing around, eating flat, round Andean bread and drinking tea from plastic mugs. They're dressed as ceremonial llama herders, with flat rectangular hats decked out with sequins and ribbons and braided

leather whips slung over their chests. Seated by the shrine is a young man with a friendly, intelligent face and a white mask pushed above his forehead.

He sets his steaming mug on the ground as we approach.

He is their majordomo, he tells us. Alejandro Espinoza Pando, age thirty-four.

His face knits in concentration as he gestures at the elements of his costume: the woolen face mask to guard against the cold, the sacred icons on his hat, the whips to perform the "River of Blood" dance to "feed" Pachamama.

When he turns around, I see a limp llama fetus hanging from his belt. An offering for the apu, he explains, to ensure the flock's fertility in the coming year.

Several hours earlier, at 3:00 a.m., he tells us, he led his comparsa on a moonlit climb to the glacier's edge. There they whirled slingshots and danced for El Señor de Qoyllur Rit'i. Nobody tripped or was killed by falling rocks—which would have been signs of the Lord's displeasure—so the mountain spirit must have been pleased. He would reward them with plentiful crops, maybe even a new house for some, says Espinoza.

I have learned from researcher Inge Bolin that being a majordomo is a big responsibility, especially in impoverished Andean districts. The majordomo not only organizes and leads the comparsa; they also cover everyone's costs for the pilgrimage: transportation, food, supplies, costumes. That could swallow up most of this man's yearly income, I am thinking. I hope the other people listed as majordomos on that flag chipped in a lot.

"Why do you do it?" I ask. "All this effort to organize the comparsa, to travel all this way. Can't you worship El Señor at your village?"

He looks over my head at Jorge and Paco and considers.

"The Lord of Qoyllur Rit'i wants us to live together on this earth peacefully," he finally says. "We travel to Qoyllur Rit'i in comparsas to learn to share, to stop being egotistical and hypocritical."

"Isn't that what you are learning with your comparsa, señora?"

7

THE NOTEBOOK

The Send-Off

July 21, 2012, 10:00 a.m.

I was working out on the elliptical machine, listening to some Jamiroquai to get my heart pumping. Through the floor-to-ceiling windows, I could see for miles along the Miraflores coastline: the bubbling fountains directly across the street, the Saturday morning crowds ambling farther along the Malecón, the paragliders with their rainbow-colored canopies floating over the cliffs that dropped to the rocky shoreline below. The elliptical canopies bobbed and swung, laughing tourists suspended below in strappy harnesses. They were like drunken seagulls, reckless and protected. I was not one of them; I had to keep my feet on the ground and trudge my way upward, one step at a time. I climbed faster, I was in the homestretch.

My phone lit up: Jorge.

"Your father had a heart attack."

The curtains were drawn, and Mahler's Seventh was playing. He was lying on top of the bed in his blue tracksuit, head resting on a pillow, eyes closed. Daisy and Iris were seated on either side of him, caressing his hands; two other aides sat at his feet. Iris's eyeliner was streaked.

"He died immediately, señora," Daisy said in a choked voice.

I lay my head on his flat chest. No sound.

John Drake in a garden in Chatou-Croissy, near Paris, 1951. Photo courtesy of Barbara Drake-Vera.

The aides were clustered around me, sobbing, the girls with their high ponytails and neon fingernail polish, stroking his legs and his hands and his bald head. None of it seemed faked.

"Adiós, Mr. Jhon," Daisy sniffled.

Dr. Rodríguez entered the room with Señora Estrella, his teddy bear face all somber. "He died instantly," he said. "When I got here, he had already been dead ten minutes. There was no reason to revive him." He and Jorge began conferring.

The bed creaked as I sat on the edge. I leaned over to kiss his forehead. It was dry and papery and already cold. He looked exhausted, like he deserved a good, long sleep.

Daddy, I thought.

It was a good way to die. In his own room, in his own bed, surrounded by four women full of *cariño.*

We arranged a funeral in the Congregational church in San Isidro. Only a handful of our people showed up: Henry and Mariella, some friends, the *enfermeras técnicas.* His casket was borne by six Black pallbearers wearing black tuxes and white cotton gloves. This was considered the correct send-off for the dead in Peru, and Afro-Peruvian families specialized in the service. The custom would have been considered racist as hell in the United States, but since we were in Peru and this was people's livelihoods, we just flowed with it. I put pictures of the funeral on Facebook and felt weird about it.

Two days after the funeral, Jorge, Henry, and I accompanied my father's body to a crematorium in the desert, an hour south of Lima, overlooking the Pacific. As I rode behind the hearse, pale sand dunes rising on either side of the highway, I thought about my father's unorthodox life journey, from Depression-era Easton to Pearl Harbor to Europe, back to America, ending his days in Pizarro's City of Kings. My father died in Latin America, like Sir Francis Drake had in 1596. I doubted we were descended from Drake's younger brother Robert. My father could never find any proof. That hadn't stopped him from being a blowhard about it his whole life.

He was a narcissist but a decent human being. I had loved him. We did right by him in the end. He just couldn't handle having me as his daughter.

Late August

Jorge and I flew with his remains to Easton and buried him next to my mother in the Drake family plot. My father's younger brother, Edwin, and about fifteen of my cousins and my father's surviving friends came to the funeral and valiantly followed us to the grave site in the brutal one hundred–degree heat. The gravediggers had dug a pit two feet wide by three feet long in the chocolate-brown dirt. I crouched down and deposited the urn with his ashes, along with his Scottish Rite fez and Masonic sash and dozens of pins—all the trappings that had propped him up in the early stages of dementia. I didn't care that I was burying trinkets made of 14-karat gold. These things belonged with him.

Uncle Ed leaned on his wooden cane and looked on stoically as the clergyman scattered handfuls of dirt into the shallow pit.

Lost and Found

August 2012, Gainesville

The fluorescent light buzzed overhead as I tugged the rusty handle to my father's storage unit. Jorge and I hadn't touched this stuff since my father left Florida, and we needed to retrieve some documents stored here to file insurance claims. After a few shoulder-wrenching heaves, the metal door rolled up.

There it was. The blue table.

On top of that: the rattan dining chairs, turned upside down; a rolled-up Persian rug that probably still harbored Krispy Kreme crumbs; some framed prints; two end tables. Stacked behind the table: cardboard boxes containing his classical records, his Modern Library books, tax returns going back to the year 1.

Jorge climbed on the table and pulled out the boxes we needed.

"You take this," he said, passing me an old Cutty Sark box.

We each grabbed a chair and started rifling through papers and expired driver's licenses and old high school yearbooks. My box contained newspaper clippings about iron lung machines, Robert Oppenheimer giving a speech at Columbia University (where my mother had been a secretary in the Physics Department in the 1950s), handwritten notes about different types of women's nylons, my father's navy service records.

Fifteen minutes later, Jorge said, "Got it," holding up the insurance forms. "What did *you* find?"

I stared at the old black notebook I had just unearthed—exhumed—the sixty-year-old notebook I never knew existed, its worn leather cover crumbling in my hands. The stained blue satin marker was opened to page 3.

"I think . . . something amazing."

That night in the hotel room, I read all forty-two handwritten pages of it— from New York to Paris to The Hague to the United States again.

On the first page, dated "24 *Dec. 1950*" and addressed to "amigo," was an inscription in Spanish from "Mildred and Francisco," friends of my parents before I was born. I googled the first line of the inscription; it was a sonnet from *Don Quixote,* the one with Peña Pobre, titled "Amadís of Gaul to Don Quixote of La Mancha." According to the site, *Amadís of Gaul* was a series of chivalric romances from medieval Spain about a brave knight of that name; the character Don Quixote admired Amadís more than any knight and modeled himself after him. (Peña Pobre, a footnote said, was a real mountain in the Castile Ranges of Spain.)

Curious, I read the translated first stanza in its entirety:

> You, who mimicked the tearful life of woe
> that I, in isolation, scorned by love,
> led on the lofty heights of Peña Pobre,
> when all my joy did shrink to penitence,
> you, to whom your eyes did give to drink
> abundant waters, though briny with salt tears,
> and, removing for your sake its min'ral wealth,
> earth did give of the earth for you to eat,
> be certain that for all eternity,
> as long, at least, as golden-haired Apollo
> drives steeds across the fourth celestial sphere,
> you will enjoy renown as a valiant knight;
> your kingdom will be first among all realms;
> and your wise chronicler, unique on earth.

Growing up, I had heard my father mention this Francisco once or twice in passing, bitterly, as someone who had been part of his "old set of friends" in New York and was a "user." He had never mentioned a Mildred.

Reading these lines, I wondered: Had this Francisco once played Amadís of Gaul to my father's Don Quixote?

This daily journal is kept as a record of my beliefs and thoughts. And if I have grown, over the years, that growth will be measured here also.—July 1951

Thus begins my father's writing journal. He was twenty-seven, a veteran of World War II, newly arrived in Paris with my mother, Ann, who was twenty-four and a secretary for the Marshall Plan. As I learned over the years, her boss had been a spy and was shot dead on assignment somewhere in Europe. All I knew was my father went abroad with her to write poems and a novel. Would this notebook help fill in the blanks?

Excerpts:

26 July 1951. Poetry writing is like basket weaving. It is not by chance that the design, the rhythms, the dove-tailing of thought into thought, form into form, the freedom, the celebration . . . the lamentation, the subjects, are so beautiful. It gives shape. It is whole.

I write here in order that that immortal sense of man—his thoughts, his ideas—may have its life!

7 August 1951. At the Rond Point (Ave. Champs-Élysées) today, we sat and rested, and Ann called out: "There's Orson Welles!" He came walking around, beside the bldg. of Figaro littéraire, tall, black-haired, straight-legged, husky and fat, dressed in a black velvet jacket with something white at the collar—half strolling, half roaming, a trace of a smile and looking all around, head up, conscious only of himself.

14 August 1951. Finishing Chapt. X this a.m. Studied the structure of my entire novel so far. Parts are beautiful. I believe in it. The theme is universal: a young man discovering himself, manhood, love, marriage, first job and the world! I still feel, at 27 yrs. and 5 mos., that I did right in leaving college, for life was stagnant there, unreal, and I wanted all of life. My destiny is written in my ideals and my heart.

When obstacles and problems seem great, let me remember them! Let my appetite be keen, my heart and spirit strong. I want to be the voice for my society, America, my age. Let me always write the truth—spare nothing—and no lies . . .

If I succeed in my ideals, it will be because of Ann.

16 August 1951. The first real friend I ever had was Francisco. May the world come to him. May we meet and talk much, often, over and over. Man needs healthy comradeships.

Several pages follow with proclamations about "art," "poetry," "the destiny of man."

9 Sept. 1951. Perhaps all artists should wear beards—little beards or big beards, but beards, shaped to fit the individual face. The people should thus be aware of the artist moving through their society: it would make the presence always felt.

Many pages with self-conscious observations about art, literature, daily scenes in Paris. The exuberance is stunning.

4 February 1952. Conventional poetry goes "out the window now." I write only what is fresh, new and vital, no retrogression. Say it in a new way, or not at all. Destroy all old conventional verse written by self, the imitation-poetic. Now start anew: create! My novel: I must learn to write good sentences.

8 March 1952. I am alone, truly alone . . . as any artist wrestling with a new form, a new expression of art. Don't expect everyone—or anyone for that matter—to understand what you are aiming at. "Where Are We Going?" my first novel . . . I write it pure.

9 March 1952. Something seems to have gone out of our lives. Ann no longer finds my writing inspired and wonderful. She has not read my last 3 chapters even. And my new poems, that's something too. I believe I am found picayune, trivial. My great thoughts, as I supposed them, were they not trivial?

10 May 1952. When you write, you must pay close attention to animal-instincts, for we humans are animals, with animal instincts more or less. . . . Parenthood is animal—but we humans carry it to the extreme. When the offspring is full grown, it too must have its freedom, not be sheltered, fawned upon.

29 May 1952. John Drake, because you are different, do you expect success: to be able to have a self-built bungalow in the country, the mountains and live a full life free from the world's materialism, politics? I wonder. Recipe for artistic success: first, create in tradition so that you sell well; then, financially free, CREATE. To know all this . . . and yet, I cannot but be true to "Where Are We Going?" I cannot write a potboiler.

17 June 1952. I have an intense dislike for Holland. The whole business weighs on me like the universe. Never have I been so repulsed by a country, a narrow, intense

nationalism that I find here. . . . I must get away. A number of wks. more, and then we are free. And my old self will return. I have lost all touch with nature here. There is no nature here, only sand and water.

1 Sept 1952. Home again! America. Looking through the stores and seeing all the mass-produced, cheap things we missed abroad. Comforts and necessities. Yet we, and Europeans, did without them. The stores, bargains and tempting knickknacks— all property to tie you down, make you still. The big department stores and the greeting cards. I was appalled!

10 Sept 1952. I wonder why men don't become nuns? It would be a good chance to hide a bald head!

12 Nov. 1952. My soul is imprisoned here. . . . All my hopes, dreams, joys, the cerebral, despairs. I've cut myself adrift from society. And society seems to prefer to have very little of me—my life, my writing, my cares—since I am no longer of it and will not accept responsibility, material possessions, a lifetime job or its nonsense.

The family is a prison, and the rearing of children—who will in turn rear children, denying themselves as their parents had done—what a squirrel's treadmill.

16 Dec. 1952. I am despondent today: out looking (again) for a job. But thank god I have my work.

16 June 1953. I live in Brooklyn Hts., and I love it. I can be away from all the falseness of New York and write. I detest Greenwich Village most of all. So cheap! So utterly false. The so-called artist crowd. I don't want to engage in all that talk talk talk, all that show and pretense. I am sure all the real artists are at home, working.

May 14, 1954. Finishing my first novel.

3 October 1954. Now living in Manhattan and saving $32.00/month rent. I was quite fatigued when I finished my novel—all those 4 or 5 months on the end, working 12, 14 or 16 hrs./day without let-up. The move was a good thing. For the past 12 wks. have been working in the apt., plastering, painting, building cabinets, shelves, fixing. My room is in shape at last—have made a big table, painted a flat blue—and within a few days will start my next novel.

10 October 1954. Novel rejected by Harcourt, Brace and Co., Harpers Prize Novel Contest, Viking, Scribner's, Putnam's (return letter: "Our sales advisors do not believe we could sell it to any great extent in the current market . . . in spite of some good qualities in it"). Balls. . . . What about England? They are interested in Amer. life over there, and perhaps I could get a year there in England out of publication?

16 November 1954. Writing the second novel, with an outline for the third taking shape. Must work fast now. Am shooting for June 1, 1955, as a deadline for completion of ms of 2nd novel. To market the books is not as important as the writing of them. Posterity will see to the former.

10 Dec. 1954. The Dr. Robert Oppenheimer lecture at Columbia Univ. Dr. O. is cer-tainly one of the most significant figures of this age, if not controversial. I was sur-prised to find him so short and so slight. His ears curled. His grey hair was clipped shorter, shorter than a teddy bear, and there was a bald spot in the back (which he touched, absentmindedly, from time to time). He was much embarrassed to find such a large audience, but he recovered his composure (after joking with an unlit cigarette, which he attempted to hand to the audience while the 1st speaker was delivering a eulogy of him) and soon began to talk about PHYSICS. *He did not talk down to the audience; rather, he lectured in such a manner and so exactingly, that the ordinary layman was soon fatigued, mentally (I speak of myself here).*

12 Dec. 1954. Have been working for Xmas season, part-time, in Gimbels Dept. Store, the toy department. A regular madhouse, what with mothers and fathers and chil-dren pushing, yelling and asking questions all the time. But interesting. It makes me appreciate the freedom and direction in my life. The people in the store, the regulars are down-trodden, submissive, scared and lost: One pities them, and how did they get so caught? And yet, they can be vicious, too. . . . The regulars of the toy dept. are a decent lot. But, again, one has little in common with them: they are either narrow and uncultured, or else they are homosexuals. The latter are certainly brilliant . . . but one steers clear.

8 February 1955. My novel (as yet unnamed) goes forward slowly. I am bringing it into focus. I hang the framework on the characterization of "Ernie Novak." But I must probe deeper than that. Say something about the arts. Probe. Plunge. Expose. But not too much commentary. Just show life as I see it. The reader must supply his own answers. If I can get him to "think," then I will have succeeded. But first I must re-read Joyce Carey's The Horse's Mouth *and E. Zola's* L'Oeuvre. *And then I'll sit at this homemade desk and hammer the thing into completion.*

16 February 1955. Rather "blue" these past two weeks, and don't know why. Is it because of the novel (over 6 wks. now without a word from Little, Brown and Co.) or because of disappointment with people; or because of the situation with society and the world in general? One thing, however, is certain: from all this new strength will come and fighting courage.

And I'm touchy with people, touchy and impatient with situations.

Many descriptions of novels he reads and movies and plays he sees in New York City. Sometimes two films in one day. The Thalia. A film series at the Museum of the City of New York. He doesn't mention having a job after his David Sedaris–like Christmas stint at Gimbels. He resolves to place his manuscript with a small publishing house.

9 April 1955. When I went to the museum today, after walking the width of Central Park, and approached the main steps in the sunlight, I noticed a group of Puerto Rican children playing there. I passed between them. There was much calling out and laughing. One little girl, she couldn't have been more than eight, said to me: "Mister, pretend please that you're our father and take us into the museum with you. They won't let us in alone." Momentarily, I was stunned. (I was conscious, too, of the sun beating down, and I felt: This isn't real, it's only a dream or a movie.) I said, at last, "Oh, I couldn't do that." And, as I looked to the doors of the museum and saw the attendant looking out at us, I continued: "That man there has already seen you playing together before I came." "He's right," another little girl said. "He couldn't be our father: he's an American." And there was a quick murmur of voices, and I said, "So long," and went inside. BUT THE INCIDENT PERSISTED IN MY MIND. *And then I was sad, for these children, only babies, already know one of the gravest misconceptions in life.*

Novel, returned from Little, Brown and Co. after nearly 3 months, has been recently submitted to the Noonday Press.

23 May 1955. Now concentrating on writing a short story about a physicist who is brought back to life with a heart-lung machine (but I must yet find a unique plot . . . something like K. Čapek would do!).

Two-year break

11 April 1957—A few words, to bring this journal up to date. From Nov. 1955 to April 1956, worked as a night manager in a nylon shop in Union Square Subway Station (5 to 10 p.m.). Around Sept. 1956, Francisco's son was born. . . . On Aug. 1, 1956, we moved to the East Side of town; spent 6 wks. painting and decorating. During all this time (Nov. 1954 to Nov. 1956), worked on books 2 and 3 of the trilogy; that project was abandoned, temporarily, to begin work on the heart-lung novella, The Iron Heart.

From Sept. 1956 to Feb. 1957, worked 4 to 12 p.m. in the Guaranty Trust Co., 140 Broadway. Clerical work. Met Ian and Betty Ballantine at Thanksgiving Day party. . . . From April 1956 to July 1956, worked for the Zero Press, gratis, on Zero *Anthology #8. Broke with Themistocles Hoetis when I "declined to invest" $5,000 in Zero Press.*

26 May 1957. Worked all day on the novella. I often think of myself as a voice in the wilderness. I have lost my true friend. Surely, I, unloved, stand alone. Have read Pär

Lagerkvist's Barabbas. *Very moved. Have read T. Wilder's* The Bridge of San Luis Rey. *Impressed. I seem to gravitate toward works of a moral nature.*

Two more years' break.

30 April 1959. To bring this book up to date, I note that I am now 35 yrs. old, have written 2 novels and the greater part of a novella, and about 40 poems. I know a good deal more of life, now. And, upon reading the earlier entries in this book I realize how immature I was but a few years ago. I feel estranged from the writer of these earlier entries; I almost do not know him. And yet, I wish that I had all or even some of his enthusiasm. For I am a little fatigued with life now. I save myself—for my writing. My writing is improving, and I can now write a decent line. To improve, at least that is something.

I am saddened by life because of my broken friendship with Francisco. It is true that only those whom we love can really hurt us. They alone have the power to inflict— through confidence—the mortal wound. Nothing is so tragic as friendship betrayed.

In Sept. 1958, we moved into our 5-room through apt. at 206 E. 73rd Street. I spent 6 months, off and on, painting and decorating. The apt. is lovely and overlooks the garden in the back. Ann is now working for Kaiser Aluminum.

2 June 1959. In my entry for 26 May 1957, I mention how I stand alone, "unloved and friendless." What I am really referring to is my lost friendship with Francisco. I feel that I have truly lost my only friend. I feel betrayed. This friendship meant so much to me. And now it is lost. Did I give too much of myself? Did I expose my Achille's Heel and was thus betrayed? No. I think not. It was F. who changed, not I. It was

he, who upon planning to leave the country, plotted to exploit and use me. (I am so sad about this. But the wound heals slowly.)

18 July 1959. Poetry's great themes: love, birth, death, the destiny of man. Poetry either celebrates or laments—but for me, poetry is a lament, the theme of lament. Wordsworth's "Ode: Intimations of Immortality," Thomas Gray's "Elegy Written in a Country Churchyard," Bryant's "Thanatopsis."

That is the last entry. I will be born less than two years later and brought home to the apartment on East Seventy-Third that he spent half a year painting and decorating. Sometime between that last entry and my birth at Flower Fifth Avenue Hospital, he will have joined the U.S. Post Office as a mail carrier.

Uncovered

I lay the notebook on the sofa and sank back into the cushions. My heart felt impossibly heavy, like a wet sandbag. He had tried so hard to become a writer. Nearly ten years. He wrote every day, all day, into the night. And in the end, nothing, other than two poems, got published.

I felt his pain like it was my own—because it was my own, because it is every writer's pain. When you first start writing, your work sucks for a thousand and one reasons, and you get many rejections, and you feel like the misunderstood artiste, and then, when you get tired of feeling sorry for yourself, you find some seasoned writers to tell you what in your manuscript works and what doesn't, and you start again. And you get better. And you get published. Eventually. You can't control the timeline, though.

My father never had the guts to seek out critiques from fellow writers. That was why he had failed. He only wanted to be admired.

He was a flaneur, a morose boulevardier who spent his days going to movies and plays and art exhibits and the Central Park Zoo, self-absorbed, cut off from everyone, even the "little Puerto Rican children" who begged him to take them into the museum. He could have tried taking them inside. What was the worst that could have happened—the guard saying no? No, that would have been too much for my father. Then the onus would have been on him to insist that he and the children be admitted together, and he lacked the stomach for that. More than anything, that was what he wanted to avoid—having to face the gap between his cowardice and his high-sounding ideals about the "brotherhood of man."

If only he could have realized it, elements of fascinating stories were all around him: his experiences at Pearl Harbor; my mother's work as a Moneypenny to Cold War spies; New York in the 1950s; the "phony beatniks" who drove him crazy at parties in Greenwich Village; the nuclear physicists at Columbia University, where my mom worked, carefully covering up traces of the Manhattan Project; Puerto Rican families moving into the West Side. Even his brief stint selling nylons in Union Square Station would have made a great premise for a noir novel. But no, he wouldn't write about that. He had to write the Great American Novel. About himself. "Leonard Allen."

I brushed my teeth that night and spat out, watching the white foam circle the drain. I had learned a lot I didn't know about my father's life in the 1950s. Several points stood out:

His first novel was titled "Where Are We Going?" Later he renamed it "Take the Dubious Road."

He never mentions the title of his second novel.
For the most part, he didn't work. My mother did.
After Paris, he barely mentions my mother.
A man named Francisco broke his heart.

In the following days and weeks, something deep inside me shifts. My father didn't write one novel in Paris and spend the rest of the 1950s working as a market researcher at Packaged Facts, as he had told me. He spent nearly ten years writing and polishing and trying to publish two novels, one of which I wasn't even sure existed. He had done this in Paris, The Hague, Allentown, Brooklyn, and Manhattan, writing continuously for up to sixteen hours a day, and all that time, my mother had supported them both. As a secretary. My loving, stoic, long-suffering mother who could write rings around my father, as her hundreds of letters showed.

No wonder he was jealous of me as a child. Less than two years after I wrote my first poem, I began getting published. At age nine, I was answering fan mail. It had felt like a natural progression to me, but to him, my success must have seemed sudden and unearned. It had undoubtedly torn him up inside, a man in his midforties, seeing a little girl in ponytails—writing at the same blue table he had built for himself, under the yellowing map of Paris, the city where his writer self first blossomed—achieve a fraction of what he had struggled for and craved for so long: recognition. I could almost understand his cruelty to me, almost. No wonder he drank, no wonder he became vicious when I turned to fiction, no wonder he was determined to shut it down, shut it down. And then, when I was twenty-two, why he had to kill the writer in me again.

It wasn't normal what my father had done to me. Only a deeply wounded person would have behaved like that. That's what he had been: wounded and bitter.

But I was not responsible for his bitterness. Something, or a series of things, had devastated him long before I ever came on the scene, probably even before he met my mother. He needed her to idolize him; when she stopped doing so, in Paris, he withdrew into himself. He cultivated that solitude for years, like the "lines" and phrases ("honey blond") he labored over in his unpublished novels.

The one person he still trusted after returning from Europe was Francisco, the man who knew his "Achille's Heel," whatever that was. I could think of a few things (the photos from Pearl Harbor, "but one stays clear"); I could prove nothing. And when Francisco "betrayed" him—how, over what, I would never know—that was the deepest cut. Two years later, he was still mourning it like the loss of a lover. He steeped himself in sourness and disappointment for the rest of his life—my idealistic, romantic, lyrical butterfly of a father, whose wings were bruised before he even crawled out of his chrysalis.

I was not the cause of his profound pain. That was all I needed to know.

This new understanding frees up places inside me. It is as if I have passed an enormous gallstone or, rather, released a sad, bitter homunculus who was crouched inside me all those years, whispering poison.

"Face it, Barbie, you haven't got it in you."

He was talking to himself.

A cluster of adobe houses and stone walls in Pacchanta, 2012. Photo by Barbara Drake-Vera.

On the Road

September 2012

"Are you well, Señora Barbara?" he asks in Spanish.

He looks over his shoulder at me, red and pink pom-poms dangling by his ears, a worried/mad look on his face, jaw thrust forward.

"Sí," I say faintly, clinging to the burro's white mane, my legs clamped around the animal's hard, round belly. *Panza de burro.*

Paco and I are on a rocky dirt road in the Quebrada de Pacchanta, a high valley ringed by mountains and studded with huge boulders and tiny stone houses thatched with sod. All around us are gray rocks draped with pale-green lichen and fields of spiky brown ichu grass and, above, a piercingly blue sky that turns black out of nowhere and drops hailstones the size of baseballs. No trees. Just a fierce, stabbing wind and chickens and dogs wandering around and the ever-present keening of traditional *huayno* music, drifting from transistor radios in the potato fields.

And towering over everything, to the south, the white-and-purple peak of Mount Ausangate—lord of the sparkling blue lakes and the flocks of llamas and alpacas and the people who live at this dazzlingly thin altitude, more than fourteen thousand feet above sea level.

"Cousin, sit up," Chata shouts in Spanish. "We are almost there." She has been strolling behind with Nati, our translator from Cusco. The two of them have become fast friends since we set out from Paco's farm in Upis three days ago, interviewing campesinos together.

"Yes," I bleat, slipping even more to my left. *Hang on,* I tell myself, *hang on until we reach Pacchanta, the land of the weavers.*

Paco spits in the tall brown ichu grass and lets the others pass until the burro and I catch up to where he is standing. He shoves another coca leaf in his mouth and smiles wearily. I know he's happy I called him to arrange this expedition after missing Qoyllur Rit'i. I know I'm paying more than we originally agreed on, but that's okay. I owe him this. Actually, I owe this trip to more than just Paco and his family and his cranky mother. I owe it to all the forgotten people who live up here, in the lap of Apu Ausangate.

Children of the dwindling glacier streams. Draped over my burro's back are two canvas packs with rice and jerky and beans that Chata and I brought. We've been handing them out at every hamlet we visit. We've emptied our own supplies of aspirin and penicillin. I wish we had brought more. Next time.

"Over there!" I hear Chata yell. She takes off down the road toward a cluster of two-story adobe buildings a quarter of a mile away. She has been studying the designs made by the Pacchanta weavers, geometric and animal forms dating to before the Incas, some woven nowhere else in Peru but in this small village. I know what her heart is feeling, the burst of a trickling stream after decades underground. The dust rises behind her, and then she doubles back, grabs Nati by the arm, and pulls her along. Their laughter echoes in the valley.

"Fuerte [strong]," says Paco, nodding at them approvingly.

I know who he's not including in that compliment.

He crosses to the other side of the burro and tugs my blanket sideways so I'm centered. The burro's ears flick back.

"Solpayki," I say, a word in Quechua that Nati taught me. *Thank you.*

"Hinallatapas." That must be *You're welcome.*

I sit up a little taller and let Paco guide the burro down the hill as the afternoon sun flecks the boulders with orange. Some dogs are yapping. A song is in my ear. A song Paco's mother sang for us yesterday morning by her solitary stone hut facing Mount Ausangate. Two white alpacas stood in a nearby corral.

She yelled at me for not bringing marmalade and didn't thank us for the rice. She sat on a wide stone in the fields, pulled down her wide, floppy hat, and pointed at the mountain, as Nati translated from Quechua to Spanish. Before, she said, Apu Ausangate was prettier, like a woman with many skirts. Now he had lost his white poncho. There wasn't enough water for the crops and animals. She was going to have to climb the mountain and make an offering. Wine and chocolate. Did I have any?

I am sorry, no.

Bah, she frowned. She pulled her hat lower against the sun. "Listen with two ears, señora. As I walk to Ausangate, I will sing a song to open the heavens."

THE NOTEBOOK | 283

Merciful Apu Ausangate, let there be rain.
The sowing dries up, the animals die.
Unleash the rain, merciful Father, and help your children.

¡Misericordia, Apu Ausangate, que haiga lluvia!
La siembra se seca, los animales se mueren.
¡Suelta la lluvia, auxilio para tus hijos!

(Quechua to Spanish translation by Nati Nuñez)

Her singsong voice floated over the fields, as harsh and piercing as the ichu grass. Dear merciful father. The pair of alpacas lifted their heads and listened.

EPILOGUE

More than a decade has passed since I traveled with Chata, Nati, and Paco to the lap of Apu Ausangate, more than fourteen thousand feet above sea level, but I can still recall the biting rush of the wind over the ichu grass, the keening of the *huayno* music, my own gasping for breath in that cold, thin air. Much has changed in Peru, politically and environmentally, since the early 2010s, but for me, one point remains an unchanging personal touchstone: The eighteen months that Jorge and I sheltered my father in Lima were the key that helped set me free from decades of hidden trauma, borne from my father's long-ago envy of me as a prodigious child poet and, later, as a young writer.

We all work around our own traumas, channeling our efforts around them like streams around unyielding, moss-covered boulders. Whatever their source or severity or date of origin, we often feel shame acknowledging them to others, but as I have found, the hurts endured in childhood shape who we become. For years, I pretended mine didn't exist, and that gave them power over me. They gnawed at my early belief in my writing abilities; they hobbled my efforts as an adult to publish my creative work, first stories and then book-length manuscripts, even when I took on editing and feature writing as a career. The peculiarity of my neuroses, coupled with my defiant choice of profession, compounded my need to stay silent. A writer who experienced getting published as being punished? It was both painful and absurd.

Then, in 2018, I showed an early draft of this book to a friend, novelist Jill Ciment, who asked, in her characteristically direct way, what kind of relationship I had had with my father over the years. I told her. "He was jealous of you as a little girl? Write about that," she said.

Putting the story down on paper completed a cycle I had begun in Peru. I was free.

While revising this book, I learned about the relatively new psychological concept of *complex trauma*. The Foundation Trust defines it as "the exposure to multiple, often interrelated forms of traumatic experiences AND the difficulties that arise as a result of adapting to or surviving these experiences." Although my brilliant and empathetic Manhattan psychotherapist did not use the term in the 1980s, I believe this is what I experienced as a young adult, curled in paralysis on the floor of my apartment, suffering crippling panic attacks when I began publishing again, unable to write with a man in the room. Complex trauma and complex post-traumatic stress disorder (CPTSD) have been the focus of several excellent nonfiction books in recent years, written by practitioners and by those who have suffered the disorder's insidious, soul-destroying effects. They, like I, know that emotional abuse causes hidden bruises that hurt as much as those that are visible. Freeing ourselves of trauma's grip is a lifelong process—if we have the courage and the opportunity to face our demons.

My unlooked-for opportunity came when I was just shy of fifty, an age not typically associated with self-liberation. Five years earlier, I had fallen in love with a sacred melting glacier in the Cordillera Vilcanota range of southern Peru, where people had been gathering since before the time of the Incas. I was obsessed with writing about it and about the rapidly accelerating effects of climate change that were dooming it and other glaciers in Peru to extinction. Qolqepunku Glacier and its annual indigenous pilgrimage had become personal to me in ways difficult to describe to anyone besides the handful of Andean studies scholars I consulted on the subject between 2006 and 2011. In January 2011, I was preparing to make my fourth pilgrimage to the glacier, determined to finally land a story assignment from one of the newspapers that had turned down my earlier climate change pitches, when I received the news no adult child of an aged parent wants to hear: my father had Alzheimer's, and my husband and I were now solely responsible for his care. What made the news especially painful was this was the man I had been running from my entire adult life, the once-adored, deeply narcissistic parent who had left me a scarred wreck about my writing. It did not feel like an opportunity at the time. It felt like a sentence of hard labor.

Living with and caring for my father in Peru was, indeed, hard emotional labor, but unlike most Americans, Jorge and I had a reprieve from the physical grind of caretaking. Thanks to the abundance of home health care services in Lima and the long-standing Latin American tradition of caring for elders under one's own roof, we were able to hire licensed health aides, *enfermeras técnicas*, to feed, clean, and tend to my father, in alternating shifts, for twenty-four hours a day, seven days a week. Together with the cook we hired to feed the aides and my father, our expenses totaled a fraction of what such personalized care would have cost in Florida or elsewhere in the United States. The aides' attentiveness and professional training meant that my crusty, World War II vet of a father was cared for with dignity for every moment of his last eighteen months on this earth. Supporting our aides' daily efforts taught me that caregiving is skilled, intensive work and everyone who provides it, including family caregivers, should be paid for it.

In 2022, more than eleven million family caregivers in the United States provided unpaid care for a loved one with dementia (for a total of more than six million patients nationwide), according to statistics from the Alzheimer's Association. These caregivers gave an estimated eighteen billion hours of care valued at nearly $340 billion; by 2050, the association estimates, these costs could rise to nearly $1 trillion. These caregivers, and all who come after them, deserve to be compensated for their protean efforts. While pharmaceutical companies continue the decades-long quest for a medical cure, these unpaid family caregivers do the lion's share of helping people with a degenerative brain disease survive to see another morning. They are frontline workers who deserve our society's respect and adequate financial compensation.

But Alma, Daisy, Señora Lucinda, and the other Peruvian health aides who tended to my father in Peru brought more than competent nursing to our lives. They brought humor, affection, and *cariño*, a Spanish word whose meaning I grew to define, during my seven years in Peru, as "caring through action." The act of being present for another's suffering and doing what we can to alleviate it is the essence of giving—an act we are obligated to perform for others and for the planet. This change in my understanding is what unlocked my own liberation, and I was nudged toward this worldview by not just the *enfermeras técnicas* but also my fellow professors at the Universidad Peruana de Ciencias Aplicadas and by the people of the high Andes whom I walked beside and befriended.

I came to Peru a self-sufficient, admittedly self-absorbed, American. I longed to awaken U.S. readers' hearts to the realities of climate change by informing them about an imperiled tropical glacier that was essential to the survival of the local people, their crops, and livestock, as well as being central to their ancient spiritual practices. My father's illness temporarily ended that quest, and when I was free to report again, I found that the wider story of Qoyllur Rit'i and climate change had been broken in my absence by other journalists, including a reporter for the *New York Times* in May 2012. (More major news outlets followed in the 2010s, including the BBC, the *Los Angele Times*, the *Washington Post*, and the *Guardian*.)

Surprisingly, being "scooped" after eighteen months' confinement in dusty desert Lima did not hurt as much as it once would have. The humbling act of taking up teaching to support my father's care had shifted my priorities. *Good*, I thought, reading those well-reported stories. *Word is getting out at last.* Finally, U.S. media outlets were pivoting to reporting on climate change, rather than providing a forum for a manufactured debate over the crisis's origins or existence. The tides were shifting, and I was gratified to know that others were now informing readers in North America. I knew Jorge and I would soon leave Peru to return to the United States, but I made sure to arrange one last pilgrimage, not to the yearly spectacle of Qoyllur Rit'i but to some of the isolated hamlets in the foothills of Mount Ausangate, where many of the pilgrims live year-round. I owed it to Paco and his people to bear witness to the loss of their pasturelands, their animals, and their connection to the life-giving glaciers. Chata and I brought food, medicine, and a voice recorder to capture their words and songs. Being there and helping those in need mattered more than a byline, I now knew.

I wrote a book about the screwed-up relationship between my father and me and how I fixed that (sort of) and fixed myself (a bit more), but perhaps the most important character in *Melted Away* is still in trouble: the tropical glaciers of the high Andes—and by extension, all the glaciers around the world.

The term *tropical glacier* sounds like an oxymoron. These glaciers form at high altitudes on mountain peaks in low-latitude regions, found between the equator (0 degrees N/S) and 30 degrees N/S. For one-sixth of the world's population living in places like South America, Africa, and Indonesia, they are a major source of water that is rapidly melting away, due to the effects of climate change. That loss threatens the millions of people who depend on them for water for drinking, farming, and electrical power generation.

High-altitude tropical glaciers are sensitive to the slightest changes in temperature and precipitation, showing signs of stress before ecosystems at lower latitudes. In the 1990s and 2000s, they were our "canaries in the coal mine," harbingers of a rapidly heating atmosphere whose cries went unheard by nearly everyone except the inhabitants of these high regions and the international climate scientists, like Lonnie Thompson, conducting research there.

In 2006, I visited farmers in Peru's Urubamba Valley who held up recently harvested ears of corn and those from the mid-1990s, comparing the two. The newer ears were about four inches shorter and an inch thinner in circumference, due to lessening meltwaters and higher temperatures. Three years later, in the mountain village of Pucarumi, a nine-year-old boy with chapped cheeks led me to a dried-up glacier stream that had sustained his village since his grandfather's time. Now this stream had stopped running, he told me. His people harvested fewer potatoes. More alpacas were dying. His parents talked of leaving the village.

These rural Peruvians were grappling with how to adapt to climate change effects in the 2000s. They did not have the luxury of haggling over whether climate change was real or a hoax, as did many Americans at that time, including a Miami TV reporter who laughed in my face in 2008 when I tried to speak to him about the impending dangers of sea level rise.

"I'll be dead and you'll be dead before that happens," he snorted. "That's not for another hundred years."

Well, neither of us has died yet, and of course, he was wrong. The effects of climate change are now wreaking havoc in Florida and North America and the rest of the world. Sea level rise is toppling beachfront properties in the

United States. Hurricanes are dumping more water, unleashing more powerful winds, and hovering longer over land, thanks to warmer air and sea surface temperatures. While global temperatures have risen about one degree Celsius (1.8 degrees Fahrenheit) in the last century, the Alps have warmed about two degrees Celsius (3.6 degrees Fahrenheit) in that time. Prickly pear cacti, which are normally found in dry, hot regions, are now proliferating in the Swiss Alps. I find that image as haunting as that of the forlorn polar bear stranded on an ice floe, the poster child of global warming in the 2010s.

The message that Peru's tropical glaciers held for other regions in the 2000s is now literally at our own doorstep. It has all happened so fast. We in the United States especially refused to listen—or take meaningful action.

Solastalgia. The grief induced by witnessing the destruction of a familiar landscape due to environmental factors beyond one's control. When one's home is no longer recognizable. When home no longer offers the comforts of home.

As a visitor to Qoyllur Rit'i—not someone with ancestral ties to the high Andes—I nevertheless felt this sharp, disorienting emotion when I revisited Qolqepunku Glacier in 2008 and saw that the ice wall I had leaned against just two years earlier was no longer there. The glacier's recession was just one expression of ongoing losses of the Andean ice caps in recent decades.

Due to climate impacts, the total area of Peru's glaciers has decreased by 43 percent in the last forty years, records show. In the Cordillera Vilcanota, the mountain range in southern Peru that is home to Mount Ausangate and Qolqepunku, the glacierized areas have shrunk 54 percent since the 1970s, according to research by Liam Taylor and his colleagues at the University of Leeds in the United Kingdom. The researchers note that 82 of 257 glaciers in the Cordillera Vilcanota have "disappeared completely" since 1975, "most notably a cluster of glaciers to the south-east of the Quelccaya Ice Cap."

Now that I have climbed these mountains and touched the glaciers at Qoyllur Rit'i and other parts of Peru, I cannot read these statistics without certain scenes springing to mind. The puddle of dripping water forming underneath a bespectacled Peruvian glaciologist, perched on Pastoruri Glacier, as the NBC News team interviewed him on camera in 2009. The worried face of an elderly woman in Upis as she gestured to the shrinking grasslands where her herd of alpacas grazed. The rusty tankers delivering overpriced drinking water to the impoverished shantytown inhabitants on the desert outskirts of Lima.

How much longer until the kitchen taps in Lima run dry, I wonder? How long until water wars erupt in the Western hemisphere?

Not everything that the climate disaster has brought to Peru is all bad, at least in the very short term.

In 2015, glacier recession at a peak in the Ausangate chain, formerly known as Mount Vinicunca, exposed mineral deposits arranged in stunning stripes of rose, turquoise, lavender, and gold. Rechristened "Rainbow Mountain," the site has become a major tourist destination in Peru, and trekking companies offer guided tours to the region, with hostels and guesthouses now dotting the parched valleys where Chata and Nati and I once walked with Paco.

At least for now, there is an alternate income stream for the children of the glacier, although I doubt this strategy can sustain them for long. It recalls an eye-popping enterprise I witnessed around fifteen years ago on a plaza in Huaraz: a pair of Peruvian entrepreneurs had set up a card table with a three-foot-long block of glacier ice and were shaving off shards to mix with flavored syrups, selling the treats to local schoolchildren.

When life gives you global warming, make fifty thousand–year–old snow cones.

There have also been serious efforts undertaken in Peru to address the effects of climate change—and environmental devastation in general—since I lived there in the early 2010s.

More international funds have been poured into cataloging Peru's glaciers and measuring the impacts of climate change. Among them, the National Geographic Society's current capstone product, "Visualizing 90 Years of Water Tower Transformation in the Peruvian Andes," focuses on the Cordillera Vilcanota. Its researchers are using Google Earth satellite images from recent decades to analyze how much of the country's glacier area has disappeared. They are also holding workshops with local inhabitants to learn of the impacts of rapid glacier loss in indigenous Quechua communities.

"The mountains give us water. Water is life," the local people tell the researchers. It is a simple equation and the basis of existence for all living things on this planet.

And late in the game, Apu Ausangate is receiving his proper due. On August 3, 2022, the National Geographic Society placed a weather station just below the summit of Mount Ausangate, at 20,830 feet above sea level, making it the highest weather station in the tropical Andes.

A year and a half earlier, in December 2019, an even more significant measure was passed. After ten years of advocacy and consensus building by various groups, the Peruvian Ministries Council approved the creation of the Ausangate Regional Conservation Area. This officially designated, protected area encompasses 165,000 acres in the Cordillera Vilcanota, including the endangered Quelccaya Ice Cap. Miguel Ángel Canal, Cusco's regional director of natural resources and environmental management, noted that the area "is considered a global thermometer where the relationship between global warming and glacier melting can be studied."

The designation also protects Ausangate from mining operations, putting a halt to pressure from mining companies that were petitioning the Peruvian government to allow concessions.

I am heartened by these developments, as limited as they are. However, I know that Peru and its people will suffer greatly in the decades ahead as global emissions continue to upset the ecological balance of this Andean nation—and all nations on our globe.

Even the most cynical climate change deniers have largely fallen silent now, having retreated to an equally cynical doomsday stance of "It's too late, there's nothing we can do; therefore, we don't have to do anything to reverse climate change."

Nonsense.

There is much that we can do to mitigate greenhouse gases and to help ourselves and others adapt to rapid climate change. The question for us in developed nations is, can we find the will to do it? Because the fate of the world largely depends on us, as it did on our forebears when they flung themselves so enthusiastically into the forward thrust of the industrial revolution.

It is beyond the scope of this book to outline how to do that. For those looking to get started, I suggest reading *The New Climate War: The Fight to Take Back Our Planet* (2021), by Michael Mann. It offers concrete steps individuals can take to force our governments and corporations to make real change.

It all starts with caring and acting on that impulse in ways that move the needle forward, no matter how small or insignificant we may feel as individuals.

Whenever I slide back into apathy, I remember a schoolgirl I met in Huaraz when Jorge and I attended a glacier conference there in 2009. She and her classmates had drawn pictures of their mountains and farms and animals that were displayed on the walls of her classroom, with their thoughts about the environment written underneath.

"Glaciers give us water." "Water is life," the hand-lettered notecards said. "How beautiful our mountains are, so pretty with their white ponchos."

This girl—she must have been eight or nine, with long, black braids—came forward excitedly when the scientists and Jorge and I toured the room. When I crouched to examine her colorful picture of a house with enormous birds flying overhead, she tapped my arm politely.

I looked up to catch a flash of anguish in her deep-brown eyes.

"Señora," she said, "what can I do to stop global warming? We are so worried."

At the time, the United States was producing, per capita, eleven times the greenhouse gas emissions of Peru. But she, a victim of climate change effects, was asking me, the American, what she could do to solve the problem.

I was stunned and could not answer.

There is no answer in words, I have since learned. There is only caring for others and the planet, expressed through heartfelt action.

In other words, *con cariño*.

ACKNOWLEDGMENTS

I stand on the shoulders of many who came before me in memoir, environmental writing, and caregiving advocacy.

My deepest appreciation to Jill Ciment, who urged me to write about my "strange" relationship with my jealous father. Jill read the book in various incarnations and gently prodded me to get to the story's heart.

Mil gracias to fellow redhead Joann Biondi, whose unwavering support and fighting spirit have seen me through decades of writing in Peru and Florida. *Melted Away* would not exist without her.

I am also grateful to Cynthia Barnett, whose writings on the environment inspired me. Her personal encouragement buoyed me on the road to publication, as has her general advice to climate change reporters to "write for the caring middle."

Heartfelt gratitude to James Long for believing in this book and for making a home for it at LSU Press. I could not have wished for a calmer, steadier hand than his. Also, many thanks to Neal Novak, Elizabeth Gratch, James Wilson, Sunny Rosen, and the rest of the LSUP staff for their consummate care with every element of this, my first book.

Like the rest of humanity, I owe an immeasurable debt to the many researchers who have been documenting climate change and indigenous cultures in Peru for decades, often under increasingly perilous conditions. I give special thanks to paleoclimatologist Lonnie Thompson for his long-standing generosity in answering my many questions via email and in person. Also, I am indebted to Mary Davis for granting permission to use the composite image blending Martín Chambi's 1935 portrait of a pilgrim at Qoyllur Rit'i with Mary's photo taken at the same site in 2006, showing the retreat of the glacier.

Anthropologist Inge Bolin lit the way with her extensive field research in the Peruvian Andes and with her comprehensive analysis of the intimate relationship between the indigenous people and the *apus*. I am sustained to this day by her interest in my nonscholarly work about Qoyllur Rit'i and by her fierce advocacy for the forgotten people of the high Andes.

Alfonsina Barrionuevo and Jorge Flores Ochoa (1935–2020) were among the seminal figures in Andean studies who graciously met and corresponded with me over the years. Their insights percolate throughout these pages.

Gratitude to Wendy Weeks for igniting my curiosity about the cosmology of Qoyllur Rit'i through the writings of her late husband, anthropologist Robert Randall.

Many thanks also to Jorge Recharte and the Instituto de Montaña for inviting me and Jorge as participants to the international conference "Adapting to a World without Glaciers," held in Lima and Huaraz in July 2009 and sponsored by USAID and Peru's Ministry of the Environment. Being part of a working group at this interdisciplinary conference affirmed that the input of writers and artists is vital to creating sustainable futures.

This book was also enriched by my three years advocating in the United States for the Alzheimer's Association, which supports those with dementia, their unpaid family caregivers, and the ongoing search for treatment and a cure. A special thanks to my fellow Floridians Michelle Branham and Evan Holler for teaching me that being an agent for change begins with the simple act of sharing my own caregiving journey.

Ideally, I would thank the glaciers of Qolqepunku and Pastoruri for having opened my heart to their immense, fragile beauty; however, they have passed the terminal stage and, like all other tropical glaciers in Peru, are doomed to extinction. Instead, I must express my gratitude to the people of Peru for teaching me the essential lesson of these times: our collective obligation to put the welfare of others and the planet above our own selfish needs.

Much love and *cariño* to Señora Lucinda, Alma, Daisy, and Maggy for caring for my father during his most difficult hours. If I was able to release long-standing resentments toward him, it was largely because of their patience and deeply competent care. I am also indebted to my father's doctors and to the staff of the *casa de reposo* in San Isidro for making his last months on earth safe and comfortable.

I also thank my directors and fellow instructors in the Professional Translation and Interpretation Program at the Universidad Peruana de Ciencias Aplicadas, in Lima. They encouraged my professional as well as personal evolution, which is chronicled in this book. *Un abrazo fuerte* to my dear friend and fellow professor Iriana Valdivia.

Above all, I would like to express my deep gratitude to Paco, his family, the remote communities of Upis and Pacchanta, and the pilgrims of Qoyllur Rit'i. These "children of the glacier" welcomed me into their daily lives and into their rituals for El Señor de Qoyllur Rit'i and Apu Ausangate. The interviews they granted formed the spine of the fieldwork chapters in this work, and I hope this book brings their reality to a wider consciousness.

Thank you to Nati Nuñez for translating the words of Quechua speakers I encountered while hiking the Ausangate circuit.

Much love to the Vera and Du Bois families in Peru for welcoming me into their lives and making a place for me at Sunday *almuerzo*. Thank you, Henry and Mariella, for your insights and humor. A special thanks to Dickie for supporting Chata's last trip with me to Ausangate.

Deepest affection to Sam Vera for cheering his mother on as she wrote this book.

Finally, and most importantly, I thank Jorge Vera Du Bois for pushing me up the mountain at Machu Picchu in 1995 and for all the mountains we have climbed thereafter. His love and steady support—along with the many cups of Assam tea he brewed—sustained me through countless nights of writing and revision. I thank him for being a trusted reader, for generously sharing his photographs in these pages, and for always believing in me.

BIBLIOGRAPHY

Albrecht, Glenn, Gina-Maree Sartore, Linda Connor, Nick Higginbotham, Sonia Free-
man, Brian Kelly, Helen Stain, Anne Tonna, and Georgia Pollard. "Solastalgia: The
Distress Caused by Environmental Change." *Australasian Psychiatry* 15, no. 1 (2007):
95–98. https://doi.org/10.1080/10398560701701288.

Álvarez Blas, José. *El Apu de las nieves: Qoyllur Rit'i y Corpus Christi / The Snowy Apu: Qoyllur Rit'i and Corpus Christi.* Prologue by Alfonsina Barrionuevo. Lima: Clinica San
Pablo, 2006.

Alzheimer's Association. *Alzheimer's Disease Facts and Figures.* Ebook. Chicago: Alzheimer's
Association, 2022. https://www.alz.org/media/documents/alzheimers-facts-and
-figures.pdf.

Bolin, Inge. *Growing Up in a Culture of Respect: Child Rearing in Highland Peru.* Austin:
University of Texas Press, 2010.

———. *Rituals of Respect: The Secret of Survival in the High Peruvian Andes.* Austin: University of Texas Press, 1998.

Bowen, Mark. *Thin Ice: Unlocking the Secrets of Climate in the World's Highest Mountains.*
New York: Henry Holt, 2006.

Bradley, Raymond S. *Global Warming and Political Intimidation: How Politicians Cracked Down
on Scientists as the Earth Heated Up.* Amherst: University of Massachusetts Press, 2011.

Burney, James. "Voyage of Francis Drake round the World." Chapter 19 in *A Chronolog-
ical History of Voyages and Discoveries in the South Sea or Pacific Ocean,* 1:305–69.
1803. Reprint, Cambridge: Cambridge University Press, 2010. https://www.google
.com/books/edition/A_Chronological_History_of_the_Discoveri/h_wIitKX-KUC
?hl=en&gbpv=1.

Carey, Mark. *In the Shadow of Melting Glaciers: Climate Change and Andean Society.* Oxford: Oxford University Press, 2010.

Cervantes, Miguel de. *Don Quixote.* Translated by Edith Grossman. New York: Harper-
Collins, 2003. Online at https://www.publiconsulting.com/wordpress/donquixoteof
lamancha/#navigation.

Comtesse, Hannah, Verena Ertl, Sophie M. C. Hengst, Rita Rosner, and Geert E. Smid. "Ecological Grief as a Response to Environmental Change: A Mental Health Risk or Functional Response?" *International Journal of Environmental Research and Public Health* 18, no. 2 (January 2021): 734. https://doi.org/10.3390/ijerph18020734.

Diamond, Jared. *Guns, Germs, and Steel: The Fates of Human Societies.* New York: W. W. Norton, 1999.

Drake, John. "Elegy III." Unpublished poem, 1947–54.

———. "Take the Dubious Road." Unpublished manuscript, 1952–59.

Flores Ochoa, Jorge A. *Pastoralists of the Andes: The Alpaca Herders of Paratía.* Philadelphia: Institute for the Study of Human Issues, 1979.

Foo, Stephanie. *What My Bones Know: A Memoir of Healing from Complex Trauma.* New York: Ballantine Books, 2022.

Heckman, Andrea M. *Woven Stories: Andean Textiles and Rituals.* Albuquerque: University of New Mexico Press, 2003.

"In Peru, Melting Glaciers Lead to Water Wars." Anne Thompson, correspondent. NBC Nightly News. December 8, 2009. https://www.nbcnews.com/video/in-peru-melting-glaciers-lead-to-water-wars-317947971939.

Mann, Michael E. *The New Climate War: The Fight to Take Back Our Planet.* New York: PublicAffairs, 2021.

Randall, Robert. "Qoyllur Rit'i, an Inca Fiesta of the Pleiades: Reflections on Time & Space in the Andean World." *Bulletin de l'Institut Français d'Études Andines* 11 (1982): 37–81. https://www.persee.fr/doc/bifea_0303-7495_1982_num_11_1_1552.

Salas Carreño, Guillermo. "Climate Change, Moral Meteorology, and Local Measures at Quyllurit'i, a High Andean Shrine." In *Understanding Climate Change through Religious Lifeworlds,* edited by David L. Haberman, 44–76. Bloomington: Indiana University Press, 2021.

Salnow, Michael J. *Pilgrims of the Andes: Regional Cults in Cusco.* Washington, DC: Smithsonian Institution Press, 1987.

Thompson, Lonnie G. "Understanding Global Climate Change: Paleoclimate Perspective from the World's Highest Mountains." *Proceedings of the American Philosophical Society* 154, no. 2 (2010): 133–57. http://www.jstor.org/stable/41000095.

Velasco, Fray Salvador. *San Martín de Porres: La vida de "Fray Escoba."* Madrid: Edibesa. 2012.